Chemical Product Design

Until recently, the chemical industry has been dominated by the manufacture of bulk commodity chemicals such as benzene, fertilizers, and
polyester. Over the past decade a significant shift has occurred. Now most
chemical companies devote their resources to the design and manufacture of specialty, high value-added chemical products such as pharmaceuticals, cosmetics, and coatings for the electronics industry. The jobs
held by chemical engineers have also changed to reflect this altered
business. However, the training of chemical engineers has remained
static, emphasizing traditional commodities.

This ground-breaking text starts to redress the balance between commodities and higher value-added products. It expands the scope of
chemical engineering design to encompass both process design and
product design. The authors set forth a four-step procedure for chemical
product design – needs, ideas, selection, and manufacture – using numerous examples from industry to illustrate the discussion. The book con-
cludes with a brief review of economic issues.

Chemical engineering students and practicing chemical engineers will
find this text an inviting introduction to chemical product design.

E. L. Cussler is Distinguished Institute Professor, Department of Chemical Engineering and Materials Science, University of Minnesota. He is a
past president of the American Institute of Chemical Engineers and the
author of *Diffusion*, published by Cambridge University Press.

G. D. Moggridge is lecturer, Department of Chemical Engineering,
University of Cambridge. His research ranges broadly from new adsorbents to the control of zebra mussels.

Chemical Product Design

E. L. Cussler
University of Minnesota

G. D. Moggridge
University of Cambridge

PUBLISHED BY THE PRESS SYNDICATE OF THE UNIVERSITY OF CAMBRIDGE
The Pitt Building, Trumpington Street, Cambridge, United Kingdom

CAMBRIDGE UNIVERSITY PRESS
The Edinburgh Building, Cambridge CB2 2RU, UK
40 West 20th Street, New York, NY 10011-4211, USA
10 Stamford Road, Oakleigh, VIC 3166, Australia
Ruiz de Alarcón 13, 28014 Madrid, Spain
Dock House, The Waterfront, Cape Town 8001, South Africa

http://www.cambridge.org

First published 2001

Printed in the United States of America

Typefaces Times Ten Roman 10/13 pt. and Gill Sans   *System* LaTeX 2_ε [TB]

A catalog record for this book is available from the British Library.

Library of Congress Cataloging-in-Publication Data
Cussler, E. L.
     Chemical product design / E. L. Cussler, G. D. Moggridge.
        p.   cm. – (Cambridge series in chemical engineering)
        ISBN 0-521-79183-9 – ISBN 0-521-79633-4 (pb)
            1. Chemical industry.   I. Moggridge, G. D.   II. Title.   III. Series.
     TP149 .C85   2001
     $660'.068'5$ – dc21   00-063069

ISBN 0 521 79183 9 hardback
ISBN 0 521 79633 4 paperback   acc✓

Contents

Preface

Since its inception in its modern form around the turn of the nineteenth century, the chemical industry has largely concerned itself with the manufacture of commodity chemicals. A commodity chemical is manufactured in very large quantities (at least 1,000 tons annually) and sold into a world market where the products of different companies are differentiated only by price; quality and composition are identical. Benzene, hydrogen, polyester, and titanium dioxide are examples. This industry has been immensely productive and successful, providing the major source of employment for trained chemists and chemical engineers as well as many from other disciplines.

We suggest in Chapter 1 that this industry had its Golden Age in a period of a couple of decades just after World War II, with growth equivalent to that of the modern software industry. Since the 1970s, market growth has slowed and companies began to concentrate on consolidation and economies of scale. In the past decade a significant shift has occurred in the chemical industry, in which most new resources are now devoted to the design and manufacture of chemical products, rather than the traditionally dominant commodity chemicals. This change is reflected in the employment of new graduates from our universities. Thirty years ago most went into commodity-based companies. Now the majority start work on chemical products.

The change of direction of the British company ICI in the late 1990s is an exemplar of this shift from commodities to chemical products. Since the 1920s the mainstay of Britain's bulk chemical industry, ICI no longer manufactures polyester, fertilizer, or ethylene; instead it now makes perfumes, flavorings, and coatings for the electronics industry.

Commodities of course continue to be made – the world needs toluene, ammonia, and methanol just as it always has. However, commodities are made by a dwindling number of ultra-efficient companies, which employ relatively few people; profits remain good, if cyclical, but no longer lead to a large-scale need for chemists and engineers. Often these companies are private, allowing them more easily to ride the trade cycles typical of commodity businesses.

So we have argued that there has been a rapid shift in focus in the chemical industry in the past decade, away from commodities and into chemical products; but what is the distinction between these two? We distinguish chemical products from commodities in three ways: quantity manufactured, value, and differentiation in the marketplace. Chemical products are produced in small quantities (usually much less than 1,000 tons per year). The archetype is the active ingredients of drugs, only a few kilograms of which may be required annually to command a market worth millions of dollars. Chemical products are high value added and sell at a high margin, whereas commodities typically sell for a few hundred dollars per ton at a marginal profit of a few percent. This high value added reflects our third difference: chemical products are differentiated in the marketplace primarily by performance and quality, rather than price. A company's advantage is often sustained by patent protection, or it may rely on trade secrets and a technological lead. This form of differentiation makes chemical products vulnerable to improved competition: patents may expire; the competition's technology may get better. As a consequence, chemical products are expected to have a short lifetime, typically a decade or less.

Clearly this is interesting, but does it affect chemists and engineers? Yes. Because of their different nature, chemical products are designed and manufactured in very different ways from commodities. They are usually made in batch, in generic equipment used for many different products. Much of the training of chemical engineers is directed at the optimization of large-scale, purpose-built plants: this is irrelevant for the type of process normally required for chemical product manufacture. Furthermore, because of the short lifetime in the market of chemical products and the premium on being first to market, chemical product industries require rapid, new designs. We argue that this has led to a cultural shift in the industry away from specialization and toward integrated design teams. Where the traditional chemical engineer has been restricted to the design, commissioning, and optimization of a plant with a predetermined purpose, the modern counterpart can expect to be involved in the design procedure right from its inception through to the marketing of the product. The process design we teach chemical engineering students seems limited in this context.

Thus the changes in the chemical industry have had a very substantial impact on the type of work an engineer or chemist can expect to do. We have just stated that much of what has been seen as the core of chemical engineering education is of only marginal relevance to the new industry. Are we therefore arguing that our discipline is now irrelevant, or that it has already made the necessary changes in response to the changing needs of industry? Our answer to both these questions is an emphatic no. We believe the chemical engineering curriculum has changed remarkably little in the past 50 years, ignoring the major changes in industry. Indeed that is why we wrote this book. However, we also believe that the core skills of chemistry and chemical engineering are ideally suited to the design of chemical products.

In his presidential address to the Institution of Chemical Engineering in 1966, P. V. Danckwerts said, "It would be a great mistake to think of the content of chemical engineering science as permanently fixed. It is likely to alter greatly over the years, in response to the changing requirements of industry and to new scientific discoveries and ideas for their application." We believe academic chemical engineering has done well at responding to the latter; computerized optimization, for example, is now a standard part of design courses, and modern equations of state are normal in the teaching of thermodynamics. However, in terms of responding to the changing requirements of industry, we believe chemical engineering education has done almost nothing. A glance at an old course syllabus or textbook or a consultation with retiring academics reveals that the basic structure of chemical engineering curricula is essentially unchanged in the past thirty years, that many parts of a 1950 course are completely familiar, and that the essential elements are recognizable even in the first systematic textbooks of the 1920s.

Does this matter? After all, the teaching of basic geometry has not changed much in the past couple of millennia, but it is as useful today as ever. We wish to emphasize that this is to a large extent true of the underlying science in chemical engineering too. The principles that have served so well in process design are applicable to the new challenges of chemical products. We hope that this has been brought out by the numerous examples in the book. Indeed, we believe that chemical engineers are uniquely well placed to exploit the new emphasis on chemical products; our discipline has always drawn together the technology of other areas into a coherent engineering subject, united by thermodynamics, heat and mass transfer, and unit operations; and this discipline is an ideal background for those involved in developing chemical products. Chemical product design offers chemical engineering education the opportunity to share in a renaissance, integrating our well-established skill set into the new challenges of the chemical industry.

We do not wish to suggest wholesale changes to the curriculum, although undoubtedly some alterations are appropriate. What we do argue is that an evolution is needed in the way in which design is taught, in order to prepare our students for the types of career many will have. Design of large-scale continuous plants is simply no longer what most engineers do. An expanded design experience should prepare students for the more diverse and flexible roles they are now expected to perform in industry.

In order to achieve this we have described a four-step design procedure: needs, ideas, selection, and manufacture. This procedure gives the book its structure. Similar procedures have long existed in mechanical engineering and business studies. We have adapted these to the needs of chemical product design; we hope that in doing so we have reflected the unique technology involved in the chemical industry. We passionately believe that in this industry design must be led by chemical technology.

Our four-step procedure can of course be no more than a template, a starting point from which to proceed. It suggests a helpful structure with which to organize design: it is a heuristic from which to start, not a description of a complete or perfect strategy. It must be adapted to individual cases – every product is unique. It will naturally be appropriate to emphasize different stages for different products. Although we have presented the design procedure as four steps in series, this too is an oversimplification. Frequently, elements of the different steps will proceed in parallel. More often than not, we will return to earlier stages in the light of later conclusions. Chemical engineers understand the value of recycle loops in enhancing efficiency; this applies to the design procedure just as it does to chemical reactors. Despite its evident simplicity, we do believe that providing an intellectual framework for design is an important and valuable contribution to both teaching and practice.

Suggestions and Acknowledgments

We believe that the material in this book can be used in three different ways. First, it can supplement a conventional, two-semester course on chemical process design. This course will normally include a project for student teams, a project that can include process design, product design, or both. We recommend that the product design material be taught after the material on process design, because we feel that students benefit from learning more quantitative process synthesis before trying to make qualitative decisions concerning products.

Second, the material in this book can also be used in a one-semester course on product design. We suggest that such a course should be about one-third lectures and two-thirds tutorials on design projects. Although the lectures should be spread throughout the semester, they should be more frequent at the beginning of the course. Finally, the material in the book can be used for a short course containing lectures alone, though we believe that this may be less effective for inexperienced students of uneven ability.

We are indebted to many who helped us write this book. We benefited from the encouragement of Professor John Bridgwater, who arranged our collaboration at the University of Cambridge, and to Keith Carpenter of Astra-Zeneca, who partially supported it. We were strongly influenced by the excellent book *Product Design* by Ulrich and Eppinger, which showed us how this subject could be effectively taught in mechanical engineering. We benefited from discussions with Professor James Wei of Princeton University and Professor Hans Wesselingh of the University of Groningen. We are grateful to Liz Thompson and Shirley Tabis, who did much of the typing. Finally we would like to thank our students, who were generously tolerant as we shaped a few slogans into an educational experience.

E. L. Cussler & G. D. Moggridge
Cambridge, U.K. 3 May, 2000

Notation

a	area per volume
A	area
c	total concentration
c_i	concentration of species "i"
$\hat{C}_p, \tilde{C}_p$	specific and molar heat capacities at constant pressure
$\hat{C}_v, \tilde{C}_v$	specific and molar heat capacities at constant volume
d	characteristic length, e.g., pipe diameter
D	diffusion coefficient
f	fraction extracted
f	friction factor
F	force
$\mathcal{F}$	Faraday's constant
g	gravitational acceleration
G	Gibbs free energy
G	gas flux, in mass per area per time
G	crystal growth rate
h	individual heat transfer coefficient
H	enthalpy
H	partition coefficient; often a Henry's law constant
H	feed flux in extraction
j_i	diffusion flux of species "i"
k, k_D	mass transfer coefficients
k, k', k''	rate constants for chemical reactions
k_B	Boltzmann's constant
k_{surface}	rate constant for a surface reaction
k_T	thermal conductivity
K	overall mass transfer coefficient
K, K'	equilibrium constants
K_a	association constant, especially for a weak acid
Kn	Knudsen number (λ/d)
l	size, for example of a turbulent eddy or an adsorbent bed
l'	length of unused bed
L	liquid flux, in mass per area per time

m	total mass transferred
m	molecular mass
m	equilibrium constant between phases
$\tilde{M}$	molecular weight
n	total number of moles
n_i	total flux of species "i"
$\bar{n}_i$	average assessment of product "i"
n_{ij}	assessment of product "i" by consumer "j"
N	particle concentration, number per volume
N_i	interfacial flux of species "i"
$\tilde{N}$	Avogadro's number
NTU	number of transfer units
p	pressure
p_i	vapor pressure of pure component "i"
pH	$-\log_{10}[\mathrm{H^+}]$, where $[\mathrm{H^+}]$ is the proton concentration
pK_a	$-\log_{10}[\mathrm{K_a}]$, where $\mathrm{K_a}$ is the association constant
P	power
q	heat flux, energy per area per time
Q	heat
r	reaction rate, moles per volume per time
r	crystal radius
R	gas constant
Re	Reynolds number (dv/v)
s_{ij}	score of factor "i" for idea "j"
S	entropy
Sh	Sherwood number (kd/D)
t	time
t_B, t_E	breakthrough time in and exhaustion time in adsorption
T	temperature
U	internal energy
U	overall heat transfer coefficient
v	velocity
V	volume
$\underline{V}_i$	partial molar volume of species "i"
W	work
W	volume of adsorbent
We	Weber number ($\rho v^2 \delta/\sigma$)
x_i	mole fraction of species "i", especially in a liquid
y_i	mole fraction of species "i", especially in a gas
z_i	electrical charge on species "i"
α	thermal diffusivity
$\dot{\gamma}$	strain rate (dv/dz)
δ	thickness, often of a thin film
δ_i	solubility parameter
ε	void fraction
η	electrochemical overpotential
λ	mean free path

μ	viscosity
μ_i	chemical potential of species "i"
ν	kinematic viscosity
Π	osmotic pressure
ρ	density
σ	collision diameter
σ	surface tension
τ	characteristic time
τ	shear stress
ϕ, ϕ_i	volume fraction of species "i"
φ	electrochemical potential
ω	angular speed
ω	activity parameter
ω_i	weighting fraction for factor "i"
Ω	correction factor in estimating thermal conductivity

An Introduction to Chemical Product Design

This chapter explains what this book is about and why its subject is important. This is a book about the design of chemical products. In our definition of chemical products, we include three categories. First, there are new specialty chemicals that provide a specific benefit. Pharmaceuticals are the obvious example. Second, there are products whose microstructure, rather than molecular structure, creates value. Paint and ice cream are examples. The third category of chemical products is devices that effect chemical change. An example is the blood oxygenator used in open-heart surgery.

The nature of chemical product design is described in Section 1.1. Product design emphasizes decisions made before those of chemical process design, a more familiar topic. Chemical product design is a response to major changes in the chemical industry that have occurred in recent decades. These changes, described in Sections 1.2 and 1.3, involve a split in the industry between manufacturers of commodity chemicals and developers of specialty chemicals and other chemical products. The former are best served by process design, and the latter by product design.

The fourth section of this chapter outlines the product design procedure that we will use in the remainder of the book. This procedure is a simplification of those already used in business development. Such a simplification clarifies the basic sequence of ideas involved. Moreover, the simple procedure allows us to consider in considerable detail the technical questions implied in specific products. This technical approach is suitable for those with formal training in engineering and chemistry, and may also be challenging for those whose training is largely in business.

1.1   What Is Chemical Product Design?

Imagine four chemically based products: an amine for scrubbing acid gases, a pollution-preventing ink, an electrode separator for high-performance batteries, and a ventilator for a well-insulated house.

These four products may seem to have nothing in common. The amine is chemically well defined: a single chemical species capable of selectively reacting with sulphur oxides. The ink is a chemical mixture, which includes both a pigment and a polydisperse polymer "resin." The electrode separator should provide a safeguard against explosion if the battery accidentally shorts out. The ventilator both provides fresh air and recovers the energy carefully conserved by insulating the house in the first place.

What these products do have in common is the procedure by which they may be designed. In each case, we begin by defining what we need. Next, we think of ideas to meet this need. We then select the best of these ideas. Finally, we decide what the product should look like and how it should be manufactured.

We define chemical product design as this entire procedure. At the start of the procedure, when we are deciding what the product should do, we expect major input from both marketing and research, as well as from engineering. By the end of the procedure, when we are focused on the manufacturing process, we expect a reduced role for marketing, and a major effort from engineering. However, we believe that the entire effort is best viewed as a whole, carried out by integrated teams drawn from marketing, research, and engineering.

We can see how product design develops by considering three of the products already mentioned in somewhat more detail. For example, for the pollution-preventing ink, our original need may be to reduce emissions of volatile solvents in the ink by 90%. Our ideas to meet this need include reformulating the polymer resin in the ink in two different ways. First, by using a polydisperse resin of broader molecular weight distribution, we can eliminate the need for volatile solvents in the ink itself. Thus there will be no emissions during printing. Second, by adding pendant carboxylic acid groups to the resin, we can make the resin not only an effective component of the ink but also an emulsifying agent in dilute base. If we wash the presses with dilute base, we can clean them without volatile solvents and without solvent-soaked shop rags. The manufacture of the new ink will be similar to that used for the existing ink.

Consider the amine for scrubbing acid gases. Current acid gas treating often uses aqueous solutions of amines, such as monoethanol amine. After these solutions absorb acid gases such as carbon dioxide and sulphur oxides, they are regenerated by heating. Though this heating gives an efficient regeneration, it can be expensive. The need is for amines that can be more easily regenerated. Our idea is to effect the regeneration with changes in pressure. We would absorb the acid gases at high pressure and regenerate the amines at low pressure, where the acid gases just bubble out of solution. In order to achieve this end we have little idea how to proceed, so we are forced to synthesize small amounts of a large number of sterically hindered amines. We will test all candidates to find the best ones. We will then manufacture the winners. Like many high value-added chemicals, these will be custom syntheses, made in batches in equipment used for a wide variety of products. This obviates the need for intensive process design in many of the chemical products that we are considering.

A third example of product design is house ventilation. Well-insulated houses are energy efficient, costing little to heat, but they may exchange air at less than one tenth of the recommended rate. To get more fresh air, we can open a window, but this sacrifices our efforts at good insulation. The need is for a fresh air exchanger that captures the heat and humidity of our snug house, but exhausts stale air, with smells and carbon dioxide. Our idea is for an exchanger for both heat and water vapour for this energy efficient house. We can manufacture this in the same way as other low cost, cross flow heat exchangers. In this example, our product is a device – not a chemical – that increases health and comfort in the house.

The designs of the ink, the acid gas absorbant, and the home ventilator are examples of the subject of this book. This subject is different from chemical process design. In process design, we normally begin by knowing what the product is, and what it is for. Most commonly, it is a commodity chemical of carefully defined purity; ethylene and benzene are good examples. This material will be sold into the existing market for such a commodity, so we will know the price we can expect. The focus of our process design will be efficient manufacture. We will usually use a continuous process, depending on optimized, dedicated equipment, which has been thoroughly energy integrated. This type of careful process design is essential in order to compete successfully in the commodity chemical business.

The chemical products discussed in this book are completely different. Their promise stems less from their efficient manufacture, and more from their special functions. They will usually be made in batch, in generic equipment; or will themselves be small pieces of equipment. Process efficiency may be less important than speed to reach the marketplace. Energy integration may be of secondary value. Indeed, most of product design may occur before manufacture is even an issue.

We believe that product design merits increased emphasis because of major changes in the chemical industry. We do not argue that the chemical engineer's concern with process design should disappear. However, we do assert that the topics we study should reflect the chemical industry of today. How this has developed is outlined in the next section.

1.2   Why Chemical Product Design Is Important

Chemical product design has become more important because of major changes in the chemical industry. To understand these changes, we will review the history of the industry, using as an example the development of synthetic textile fibers. We also need to examine how these changes have affected employment.

CHANGES IN THE CHEMICAL INDUSTRY

From 1950 to 1970 the chemical industry produced ever increasing amounts of synthetic textile fibers, as shown in Table 1.2–1. Over these decades, while the production of natural fibers was about constant, the production of synthetics grew 20% per year. This growth was like that of the software industry today; Du Pont can be seen as the Microsoft of the 1950s. This was a golden age for chemicals.

However, from 1970 to 1990, synthetic textile fibers grew by less than 5% per year, about the same as the growth of the world population. From 1970 to 1990, the

TABLE 1.2–1   Growth of Textile Fibers (10^6 lbs/year)

Fiber	1948	1969	1989
Cotton, wool	4353	4285	4794
Synthetics	92	3480	8612

Note: From 1950 to 1970, synthetic fibers grew about 20% per year. Since then, their growth is 5% per year (source: Spitz (1988); U.S. Department of Commerce).

industry stayed profitable by using larger and larger facilities. Bigger profits came from consolidating production into bigger plants, designed for greater efficiency in making one particular product. Interest in computer-optimized design was a consequence of this consolidation. Such optimization meant small producers were forced out. For example, the number of companies making vinyl chloride in the USA shrank from twelve in 1964 to only six in 1972 (Spitz, 1988).

More recently, the industry has required new strategies to stay profitable. These strategies often centered on restructuring, which was three times more likely to affect engineers than the general working population. Whether called restructuring, downsizing, rightsizing, or rationalization, the strategy meant many midcareer engineers were suddenly looking for new jobs. The Engineering Workforce Commission in the USA now feels that engineers will average seven different jobs per career, a dramatic change from two per career in the recent past (Ellis, 1994). Middle management, that traditional goal of bright but not brilliant students, is no longer a safe haven. While starting salaries remain high, the envy of other technical professions, these salaries have not increased faster than average wage inflation in 30 years. In this environment, professional organizations such as the American Institute of Chemical Engineers now provide more help in job transitions and financial planning. Such organizations can no longer behave only as nineteenth-century-style learned societies.

Having exhausted optimization and restructuring as ways to stay profitable, chemical companies now have three remaining options. First, they can leave the chemical business. This option seems reasonable to a surprising number, including many petrochemical businesses. Second, chemical companies can focus exclusively on commodities. This seems a preferred strategy for some private companies, who may be better able to handle the ebb and flow of the profits from a commodity business. It implies a ruthless minimization of research and a concentration on in-house efficiency.

The third strategy open to these chemical companies is to focus their growth on specialty chemicals. Such chemicals, produced in much smaller volumes than commodities, typically have much higher added value as well. These higher added values mean that more research and higher profits are possible. Not surprisingly, many chemical companies are turning their focus to specialty chemicals.

Interestingly, this new focus has not changed the skills that companies demand from chemists and chemical engineers, though it has changed the jobs that they do.

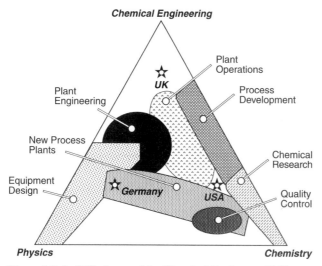

Figure 1.2–1. Skills Learned by Chemical Engineers. Skills traditionally learned by chemical engineers are a blend of chemical engineering, chemistry, and physics (including mechanics). These skills do not need major changes to be valuable for products.

The various subjects that chemical engineers learn can be positioned on the triangular diagram in Figure 1.2–1. The three corners of this plot represent training in physical sciences, in the chemical sciences, and in chemical engineering subjects. Different jobs use these three elements in different proportions, as shown in the figure. There is no surprise in this: plant engineering will demand a greater knowledge of mechanics and a smaller background in chemistry than in research and development. Figure 1.2–1 also suggests national averages. British chemical engineers have more chemical engineering and less chemistry than their counterparts elsewhere. Please do not take this diagram too literally; use it instead as a catalyst for thought. We would maintain that the basic skills needed by chemical engineers have always been diverse and have not altered dramatically.

CHANGES IN EMPLOYMENT

Although changes in the chemical industry may not have changed the skills needed, the focus of chemical companies on specialties has had a major impact on the jobs that chemists and chemical engineers do. To examine this impact, we compare the jobs taken by recent graduates with those taken by graduates 25 years ago. Our data for this are fragmentary, taken from records of graduates from the Universities of Cambridge and Minnesota. They are probably biased toward large corporations, about whom our university placement offices have better records.

The available data suggest major changes, as shown in Figure 1.2–2. In 1975, three quarters of chemical engineering graduates went to work in the commodity chemicals business. The small number who did not were split between work

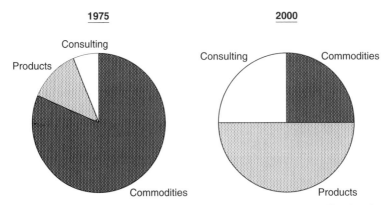

Figure 1.2–2. Changes in Employment. The dominance of commodity chemicals in the past has been eroded by the newer emphasis on products, including specialty chemicals.

on products, either design or development; and work in other areas, which for convenience we have labeled "consulting." This category includes those working directly for consulting firms as well as those carrying out specific tasks such as environmental impact statements.

More recently, the distribution of jobs has become completely different. The largest group of chemical engineering graduates, in Minnesota's case more than half, now work primarily on products. This includes students who work on materials, on coatings, on adhesives, and on specialty chemicals. The number who work in commodity chemicals has dropped so that it now is less than a quarter of new graduates. The number who work in consulting has risen dramatically, as commodity chemical businesses outsource many of the functions that they used to do in house. In one case, a commodity chemical company has taken its process engineering group from 1500 persons to fewer than 50. This is not a business cycle; this is a change in the way in which that company expects to do business; they will buy the process engineering they need from consultants. This is why the number of people involved in consulting has risen.

The emergence of products as a focus for chemical engineers implies changes in what chemical engineers do. In the past, we chemical engineers could limit our thinking to reaction engineering and unit operations, waiting for the marketing division to tell us what chemicals needed to be made, and in what amounts. Such intellectual isolation is no longer possible.

1.3   Changes in Corporate Culture

At the same time, there have been changes in corporate culture, in the ways in which all companies do business. These are in addition to the changes in the chemical industry discussed in the previous section, and they are at least as important, because they alter the ways chemical engineers work. Two major changes are especially important: the way in which corporations organize their product design, and the ways in which corporate strategy affects jobs. Each is discussed below.

CORPORATE ORGANIZATION

The organization of product development is most easily discussed by a comparison of two limiting cases: organization by function, and organization by project. These are shown schematically in Figure 1.3–1. Both can be effective.

In a functional organization, different divisions have different responsibilities: marketing, research and development, engineering, legal affairs, and so on. Product development proceeds by each division doing its job, and then passing its results on to the next division. The result is like chemical reactions in series, as suggested by Figure 1.3–1. This organization is especially associated with large, established industrial companies that have major capital investments in manufacturing. For example, the marketing department of an automobile company could discover that consumers want better climate control, that is, better heating and air-conditioning. Marketing would report their results to research, who would develop the electronic controls required for this goal. Engineering would extend the research results so that the new controls could be manufactured cheaply and efficiently. Throughout the process of product design, the development is sequential: marketing talks largely to research, only rarely to engineering. Such a functional organization can be effective, but it is almost always slow.

A common alternative is a project organization. In a project organization, a core team is formed from the different divisions. The team will normally include representatives from marketing, research and development, engineering, production, and so on. These core team members will have complete responsibility – and a good deal of resources – to design and develop the target product. They will be judged not by their immediate functional supervisors, but rather by a panel of senior managers well versed in the company's long-term strategy. Functional supervisors still have the job of making the divisions run smoothly. Such divided management can be chaotic and inefficient. As Figure 1.3–1 suggests, it is like parallel chemical reactions, with a chance of synergy between functions. Above all, this form of product development is fast, and fast product development is believed to maximize profits, so that project management is currently the organization urged by most business consultants.

Figure 1.3–1. Two Limiting Types of Corporate Organization. The project organization is currently preferred because its speed and synergism outweigh its managerial complexity.

CORPORATE STRATEGY

Superimposed on its organization, a corporation will have strategic forces driving product development. Again, the driving forces are most easily described in terms of two limiting cases. First, corporations that look toward their markets for inspiration are said to be "market pull." Corporations that emphasize extending their service and technology are said to be "technology push." Examples of market-pull companies are common. W. M. Kellogg, the manufacturer of breakfast cereal, is interested in new products from grain. The company constantly assesses the market for consumer wishes for new cereals or new grain-based snack foods. Honeywell makes home thermostats, a major product because a significant fraction of the world's energy consumption goes for domestic heating. Honeywell is interested in any new products for home comfort that can complement their thermostats. Patagonia, the maker of technical mountain climbing equipment, now also makes raincoats. This organization is trying to expand its market: many more people need raincoats than ice axes.

Examples of technology-push companies are less common but can nonetheless be found among everyday names. W. L. Gore makes Goretex, that breathable film basic to high-quality raincoats. But Gore does not make raincoats: instead, this company has used its basic material to make medical products, including arterial transplants. Exxon-Mobil has used its knowledge of petrochemical reactions to develop a series of new metallocene catalysts for polyolefins. Astra-Zeneca has used its experience with injectable therapeutics to develop delivery systems for different drugs. Interestingly, both market-pull and technology-push companies can use the same product development procedure. This procedure is described next.

1.4   The Product Design Procedure

Product design is a major topic both in subjects such as sales and marketing, and in technical professions such as mechanical engineering. Not surprisingly, schemes for this design procedure vary widely. Many are complex, especially with respect to the role of management. Many have features that seem specific to the particular subdiscipline that they represent.

The product design procedure used in this book is a simplification and generalization of those used in these other areas. It depends on four steps:

1. *Needs.* What needs should the product fulfill?
2. *Ideas.* What different products could satisfy these needs?
3. *Selection.* Which ideas are the most promising?
4. *Manufacture.* How can we make the product in commercial quantities?

The characteristics of this approach are discussed in the rest of this section. We shall see as we go along that the application of this template to the case of chemical products leads to new features of the design process.

HOW THE PROCEDURE ORGANIZES THIS BOOK

These four steps are the key to the organization of this book. Assessment of needs, the subject of Chapter 2, includes deciding on a standard for comparison – a benchmark – and converting the qualitative needs to quantitative specifications. The benchmark chosen may be an existing product or an ideal. It must be as well defined as possible so that any specifications are definitive.

Finding ideas that might meet these needs is the next step in product design. Normally, we will wish to search for a large number of these ideas by all reasonable means. This search, the subject of Chapter 3, may include brainstorming by individuals and teams and synthesizing tangent compounds by combinatorial chemistry. Once these numerous ideas are identified, they must be screened by using objective and subjective judgments, also described in Chapter 3.

At this point, we should have reduced the large number of fragmentary ideas for products to a short list of promising candidates. Typically, this reduction will be about a factor of twenty: if we start with a hundred ideas, we should have about five survivors. We must now select the best one or two for further design and development. If the characteristics of each of the surviving ideas were directly comparable, this would be easy. They normally are not. For example, we might be sure that one idea will work well but be expensive; a competing idea might be cheap but we may be unsure if it will work. Deciding between these ideas includes risk management, as described in Chapter 4.

Finally, we must manufacture the product and estimate the costs involved. These efforts, described in Chapters 5, 6, and 7, are different from those expected for commodity chemicals, where we expect to use dedicated, optimized equipment that operates continuously. Here, we will normally use generic equipment, run in batch for a variety of specialty products.

This is different from traditional chemical engineering – and is exciting.

LIMITATIONS OF THE PROCEDURE

The four-step procedure outlined above is controversial. We should review the controversies now, so that we are prepared for exceptions and diversions in the practice of product design. The controversies cluster around three criticisms: that the procedure is not general, that management and not technology is the key, and that product design is part of chemical process design. Each controversy merits discussion; they are tackled below.

First, is the four-step procedure as outlined general? It is clearly a major simplification. Many business texts argue that such a procedure is universally applicable for any product in any industry. These texts are frequently written by business consultants eager to make money by applying their own standard template to specific problems. At the same time, many professional product developers argue that this or any procedure cannot represent the peculiarities of their own industry, that only those with particular interests can hope to be effective. Although there is clearly some truth in this argument, these product developers may be like those who have denied that correlations of heat transfer could be used for food products because they were based on measurements for petrochemicals.

We believe that both sides of the debate have their merits. The four-step procedure used here is unquestionably an approximation. Certain techniques introduced in particular steps of the procedure can have value at other stages. For example, risk management, introduced in the selection step in Chapter 4, may have value in screening product ideas, explored in Chapter 3. It is unlikely that real product design will always be a simple sequential procedure as we suggest; iteration between stages is almost certain to be necessary. Still, we must begin somewhere, and the current procedure has been for us a sound and creative start. We suggest trying it; any necessary modifications quickly become obvious in specific cases. A framework in which the subject may be understood is an aid to learning in chemical product design just as an analagous template has been successfully applied for years in process design.

The second controversy is the claim that management, not technology, is key to product design. An irritating feature of most business books on product design is the extreme emphasis on the central role of management. The implication is that technology is always available if only the managers do their job properly (or at least do what the consultants say). These books on product design know no inconvenient constraints such as the second law of thermodynamics or the difference between mass and moles.

We believe that the application of technology is central to chemical product design. Product design governed only by management reminds us of a Sidney Harris cartoon showing a few managers and an engineer standing in front of a flip chart. Though the flip chart is covered with equations, pie charts, and organization charts, the engineer is pointing to one small box, which says:

"then a miracle occurs."

The engineer remarks:

"I'm having trouble with this part."

On reading books on the management of product design, we can feel all too much like that engineer in the cartoon. In this book, we want to make sure that technology is carefully considered.

The third controversy about this book is the assertion that the subject is already covered as part of the existing study of process design. This serious assertion is most easily tested by comparing our template for product design with an example of the intellectual hierarchy suggested for process design. One successful and powerful hierarchy, suggested by Douglas (1988), is summarized on the left side of Table 1.4–1. After deciding whether a process is batch or continuous, one then moves on to flow sheets of inputs and outputs which are almost always continuous. The initial flow sheets center on the stoichiometry. The next level of the hierarchy, which adds the recycles, often involves a discussion of the chemical reactions. Once these are established, one moves on to the separation trains and finally to the heat integration. All of this makes for a good course.

TABLE 1.4–1   **Process Design vs. Product Design**

Process Design	Product Design
1. Batch vs. Continuous Process	1. Identify Customer Needs
2. Inputs and Outputs	2. Generate Ideas to Meet Needs
3. Reactors and Recycles	3. Select among Ideas
4. Separations and Heat Integration	4. Manufacture Product

Note: All four steps of process design are contained in step four of product design.

If we want to emphasize product design, we need to go beyond this hierarchy. We cannot simply substitute the search for a product for drug delivery into the process design hierarchy. Instead, the four-step hierarchy suggested earlier is shown on the right side of Table 1.4–1. After first identifying a corporate need, one generates ideas to fill this need. One then compares these alternatives and finally decides on a scheme for manufacture. The manufacturing includes all of the process design hierarchy.

Thus the important steps in product design anticipate those in process design. Product design implies a focus on the initial decisions around the choice of the product and implicitly de-emphasizes its manufacture. This shifts our efforts away from the common engineering calculations that have been our bread and butter for decades. The new emphasis includes subjects that have normally been left to those directly concerned with business. It is this combination of business and technology that is the subject of this book.

1.5   Conclusions

Product design is the procedure by which customer needs are translated into commercial products. This procedure, which precedes process design, is especially valuable for specialty chemicals. Such specialties are an important focus of the present-day chemical industry, which is evolving beyond commodities that have been the emphasis in recent decades.

In this book, the product design procedure is organized as four sequential steps. The first step, described in Chapter 2, is the identification of customer needs and the translation of the needs into product specifications. The second step, in Chapter 3, describes generating and winnowing ideas to fill these needs. In the third step, in Chapter 4, the best ideas are selected for commercial development. The last step, in Chapters 5–7, includes manufacture and economics. The result is a template for chemical product design.

We must stress that management, especially senior management, is much more likely to be involved in product design than in process design. As a member of a product team, each engineer or chemist will be involved in a management review at each stage of the design process. This review will be critical; that is, the review will decide on whether the project should continue. To reflect this, in the "conclusions" section of each chapter, we will mention this review, and discuss what human interactions are likely. These human interactions are as important as the technology.

FURTHER READING

Blessing, L. T. M. (1994). A *Process-Based Approach to Computer-Supported Engineering Design*. PhD Thesis, University of Twente, ISBN 0952350408.

Cooper, R. G. (1993). *Winning at New Products, Accelerating the Process from Idea to Launch*, 2nd ed. Addison-Wesley, Reading, MA, ISBN 0201563819.

Douglas, J. M. (1988). *Conceptual Design of Chemical Processes*. McGraw-Hill, New York, ISBN 0070177627.

Graedel, T. E. and Allenby, B. R. (1996). *Design for Environment*. Prentice-Hall, Upper Saddle River, NJ, ISBN 0135316820.

Kanter, R. M., Kao, J., and Wiersema, F. (eds.) (1997). *Innovation Breakthrough Thinking at 3M, DuPont, GE, Pfizer, and Rubbermaid*. Harper Collins, New York, ISBN 088730771X.

Pahl, G. and Beitz, W. (1996). *Engineering Design, a Systematic Approach*, 2nd ed. Springer, New York, ISBN 3540199179.

Spitz, P. (1988). *Petrochemicals: The Rise of an Industry*. Wiley, New York, ISBN 0471859850.

Ulrich, K. T. and Eppinger, S. D. (2000). *Product Design and Development*, 2nd ed. McGraw-Hill, New York, ISBN 0071169938.

2

Needs

Chemical product design begins by identifying customer needs, those unfilled wants that are the original spark for product development. Who these customers are and how their needs can be identified is the subject of Section 2.1. Sometimes our product will be used primarily by consumers; in this case the needs will often be described in nonscientific terms that are hard to quantify. In Section 2.2, we explore special problems associated with these consumer products, where we may wish to use subjectively assessed product attributes.

Needs are often vague, qualitative desires for solutions to ill-defined problems. To make these needs more definite, we can set particular specifications that our product must meet. Setting these specifications is explained in Section 2.3. Usually, the specifications will require continuing revision and re-evaluation. This revision can be greatly facilitated by using "benchmarks," which are often competing products that we hope to replace. The revision of specifications and the use of benchmarks are the topics covered in Section 2.4. At the end of the chapter, we will have a basis from which to begin designing a successful product.

2.1 Customer Needs

Elucidating customer needs involves three sequential steps: interviewing customers, interpreting their expressed needs, and translating these needs into product specifications. In each of these steps, we must be careful not to narrow the product definition prematurely. We will have to resist both ourselves and our colleagues, who will immediately identify potential new products and see some good ways to improve existing products. However, at this stage, we want to focus on stating a specific need, not on meeting that need.

INTERVIEWING CUSTOMERS

Our primary source in identifying customer needs should be the final users of the product. These users may not be those who buy the product from us; rather, it is those who will actually benefit from its chemistry. These users may not be individuals, but can be organizations, including government agencies. Identifying these users, contacting them, and arranging times to discuss their needs is normally straightforward.

One group of customers are special and merit extra attention. These are the so-called "lead users," who depend very much on existing and competing products. These lead users will have needs that are well in advance of the marketplace. Accordingly, they will benefit most from any product improvements. Lead users are especially important in identifying needs for two reasons, both of which reflect their deeper experience. First, they often invent minor product improvements on their own. Second, they can often clearly express what is wrong with existing products. When we can identify such lead users, we will find them an invaluable source of information.

The consensus in the business press is that face-to-face interviews are the best way to discuss needs. Studies suggest that fewer than ten interviews may miss important information and that more than fifty result in little new information. We suggest the normal target should be about fifteen. In some cases, when you expect only one or two corporate customers, getting this many interviews may seem silly. We strongly urge talking separately to many persons within a single corporate customer, for this often exposes different opinions that shape our product design.

Two alternatives to individual customer interviews are focus groups and trained test panels. Focus groups normally have a leader and perhaps eight panel members. Their discussions, which are sometimes videotaped, can supply a synergism leading to suggested innovations. Focus groups frequently show a smaller variance of opinion than individuals. However, many in market research seem unconvinced that they are superior to individual interviews.

Trained test panels are common in evaluating small differences in consumer goods. In these cases, the panels may be encouraged to use words with very specific definitions that can differ from those in popular use. The panels can guide consumer product improvements, but we believe that they are usually less useful for chemical products, where the primary customer is often not a consumer.

Before beginning the interviews, we need to decide what we want to obtain from them, both in terms of the product and for the benefit of our organization. This is best accomplished by having each member of the core team write out the project scope, the product's target market, and the key business goals. The project scope should be one, simple sentence. The target market should include both the primary and secondary customers, and estimates of the sales that can be expected from each. The business goals should include both the timing of the new product, and the organization's technical advantage in such an effort. After each member of the core team has completed this assignment, the entire team should reach a consensus for each item. This may seem silly; after all, every team member should know this information. In our experience, they will not. We are routinely astonished at the ignorance of engineers and scientists about business and the parallel ignorance of marketers about technology. For example, in the drug industry, development chemists may not know the value of a new antibiotic; and those in marketing may not understand the cost savings possible with whole broth extraction.

Once this consensus is reached, the core team should decide on a standard interview format to ensure a common starting point. Often effective questioning can start with a list:

What do you do now?
How do you use the existing product?
What features work?
What does not work?
How do you buy the product?

We recommend that this list be simple and generic, without references to specific product ideas.

The interviews themselves depend on the interviewers' skill in eliciting useful responses. In many cases, the core team will want to seek extra help, most commonly from marketing. If possible, however, all core team members should participate in at least one interview. Members who do not do so tend to become passive, leaping into life only when they feel their own expertise is critical. In every interview, we will want to observe the following guidelines:

1. *Encourage tangents.* These will force your thinking beyond the confines of your preconceptions.
2. *Stimulate with alternatives.* Interviews often start fast and then stall. Have some alternatives, illustrated if possible, ready to restart discussion.
3. *Remove assumptions.* Often existing product use will be constrained, for example by cost or by a particular temperature range. Ask for responses when these constraints are removed.
4. *Be alert for surprises.* Customers will often describe their dreams only after they urge extending existing products. These dreams offer chances of real innovations.

The result of the interviews will be a potpourri of responses, most of which will be incomplete. We now need to organize these.

INTERPRETING CUSTOMER NEEDS

The customers' needs recorded in the interviews will normally be a random collection, filled with redundancy and irrelevancy. Our challenge is to organize these needs as groups, and to edit them into a cogent list. In this effort, we may decide to drop stated needs, even when these are suggested by many separate interviews. For example, those interviewed may ask for what we believe is a perpetual motion machine or for a product well beyond our company's anticipated expertise. Dropping these needs is fine. However, we want to do so consciously, carefully remembering our assumptions to ensure that these omissions are appropriate.

We next want to rank our list of needs as essential, desirable, and useful. We should believe that the new product must meet all the essential needs to be successful. We would like it to meet many of the desirable needs, especially if existing competitive products do not meet these needs. We acknowledge the existence of the useful needs, though we are not planning to design products explicitly to meet them. In many ways, the system is like that pioneered by the Michelin Travel Guides for ranking tourist sites. Sites with three stars are "worth a journey," sites with two stars are "worth a detour," and sites with one star are "interesting." We want to rank our needs to make our design journey profitable.

The ways in which we organize and rank needs differs depending on whether we seek to improve an existing product or to invent a new product. If we seek an improved product, we will know how the existing product is used. We will be able to define the essential, desirable, and useful product attributes without much effort. In doing this, we want to begin with the core team, discussing the needs until the team reaches consensus. Often, the team will decide that their ranking of needs requires an additional review with customers, especially with lead users. The degree to which this additional review is important will depend on how major the improvements are compared with the existing product. If we are seeking a new product, we may simply be forced to group the ideas by target market or by common function. For a new product, we almost certainly must return to the customers, perhaps a somewhat different group of customers, seeking to specify needs more tightly.

We can best illustrate these general ideas with specific examples.

EXAMPLE 2.1–1   BETTER THERMOPANE WINDOWS

In hot or cold climates, windows must provide insulation as well as allow the entrance of light. To provide this insulation, windows often consist of two or more panes of glass, narrowly separated by a gap, as shown in Figure 2.1–1. Ideally, this gap would be evacuated, but the vacuum can pull the panes of glass together.

As a result, the gap between the windows is filled with a gas and glued with a sealant. The gas should have a low thermal conductivity to be an effective insulator. Because thermal conductivity in a gas is proportional to the inverse square root of the molecular weight, the best gas would have the highest molecular weight. Historically, freons such as CCl_2F_2 (mol. wt. 121) were used. Since freons have

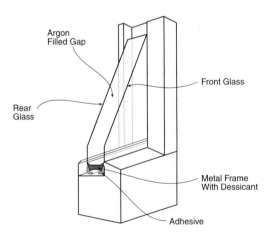

Figure 2.1–1. A Thermopane Window. The space between the panes is now filled with argon, a replacement for freon. The argon can leak out and water vapor can leak in; each can cause problems.

been phased out, argon (mol. wt. 40) is the industry's choice, even though its thermal conductivity won't be that much different from that of air.

However, the problem with the windows is less the insulation that they provide than their appearance. This appearance relies on the sealant used to glue the glass panes together. Water vapor diffuses through this sealant into the gas gap between the window's panes. This water vapor can then condense when the weather cools, fogging the window. To avoid this, manufacturers put a dessicant between the window panes, but this becomes saturated with time. In addition, the argon can leak out faster than air leaks in, causing the panes to bend together. Invent a better window sealant.

SOLUTION

This example is unusually straightforward. The most important customers for this sealant are the manufacturers of thermopane windows, whom you definitely will want to interview. You may also want to talk to some building contractors to discover if window fogging depends on how the window is installed or where in the house it is located. You probably will gain less from talking with homeowners, even though they are the eventual buyers of the windows.

The key questions you want to ask are also straightforward:

1. What are the requirements of the sealant?
2. What sealants have already been tried?
3. What is the temperature change that the window encounters?
4. What is the window lifetime?

Although there are many other questions which are helpful, they are not as central.

The answers to these questions are unambiguous. The sealant has two main functions: sealing the window panes together, and keeping argon in and water vapor out. The first of these requirements is essential; the second is that which can compromise current thermopane windows. While over 100 sealants have been carefully tried, silicone rubber gives by far the best seal. However, it is highly permeable to water vapor: the permeability of water is around 4×10^{-4} cm^2/sec, about forty times faster than diffusion coefficients in normal liquids. One commonly used alternative to silicone rubber is polyisobutylene, because its permeability to water is about 400 times less than silicone rubber. However, polyisobutylene does not make as good a seal: it does not stick to the glass walls.

Questions 3 and 4 are important to window manufacturers, but have less impact on the choice of a sealant. Most windows encounter diurnal temperature changes of less than 15°C. Despite this, in an effort to test their products under higher stress, window manufacturers can use swings of three or more times this. Window lifetimes are normally expected to be more than twenty years. However, because few people own their homes this long, the key may be to avoid water condensation for only as long as the average homeowner stays in that particular house.

The need is thus for a sealant that bonds as well as silicone rubber, but that has a water permeability at least as small as polyisobutylene.

Figure 2.1–2. Deicing Aircraft. Snow is removed by spraying the planes with a 70°C aqueous solution of 50 wt. % ethylene glycol. We seek an alternative deicer that can be recycled.

EXAMPLE 2.1–2 ALTERNATIVE FLUIDS FOR DEICING AIRPLANES

Minneapolis–St. Paul has a major airport with over 400 flights daily. This high traffic reflects the use of the airport as a hub by a major airline. This city is infamous for its cold winters. In the winter, snow can collect on planes as they wait at the gate for take-off. As shown in Figure 2.1–2, the snow is removed by spraying the planes with deicing fluids, which are discarded after use. These fluids are often discharged directly into groundwater, even though they can be toxic to humans and wildlife. In major airports like Minneapolis–St. Paul, the discharged deicing fluid is sewered, causing a major burden on the local sewage treatment plant. These deicing fluids cause major pollution.

Find alternative deicing fluids that are environmentally less abusive because they are easily recycled.

SOLUTION

In this case, we already have a product that works well, and we seek alternatives that not only work well but cause less pollution. Interestingly, our customers are not the airlines but the airports. In practice, the airports contract deicing to local engineering firms. We should interview the engineering firms' employees who are directly responsible for the deicing.

A synopsis of these interviews is given in Table 2.1–1, following the format suggested above. The synopsis does not contain some important information that does not easily fit into this format. First, all airports and all airlines are worried about this pollution problem. Some airlines are considering a special hanger with infrared heaters, an interesting idea for airplanes fully loaded with fuel! While North American airlines insist on virgin ethylene glycol, Dutch airports distill the effluent and recycle the glycol.

We next want to extract from the interviews the key characteristics that a perfect product would have. In this case, we conclude that the perfect product

1. is sprayable;
2. has a low volatility;

TABLE 2.1–1   Interview Results for Deicing Fluids

What do you do now?
"We no longer deice at the gate, because this was too difficult to control. Instead, when they are ready for take-off, aircraft are moved to a central location for deicing. We collect the run-off from the deicing in underground tanks and then slowly bleed it off to the sewage treatment plant."

How do you use the product?
"We spray with a 70°C solution of 50% water and 50% ethylene glycol ($HOCH_2CH_2OH$). We spray for 10 minutes or till there is no snow visible, whichever is longer. We then spray with anti-icing fluids." (These are snow-melting hydrogels that adhere to the aircraft while it is waiting to take off, but are removed by shear during take-off. In this example, we ignore anti-icing fluids.)

What features of the product work?
"It's a good product. It works even at −30°C. It isn't volatile. It doesn't cause corrosion, like salt. It's hard to burn." (Once burning, it has a rate of combustion per kilogram around one third that of ethane.)

What features of the product don't work?
"None. It has some odor, and some passengers get sick, but not many. The effluent contains about 2% ethylene glycol, which is toxic to fish. The environmental agencies say it is probably toxic to humans, but I'm not convinced it is. Still, they won't let us discharge it, and they're always threatening to shut us down."

How do you buy the product?
"We get it through the State of Minnesota, who require bids. One company has the largest share but we always have at least two suppliers to ensure some competition."

Note: This is a synopsis of conversations with engineers responsible for deicing aircraft.

---

3.   does not smell;
4.   is not toxic to fish;
5.   is not toxic to humans; and
6.   is easily recycled.

These characteristics imply no surprises.

However, Table 2.1–1 also shows that the interviews contain contradictions and redundancies. We want to remove these in our list of characteristics of a perfect product. For example, the product cannot be nonvolatile and still smell. The smell makes some passengers sick, so the product is volatile. As a second example, the product should be nontoxic to fish, humans, and everything else. We do not need to list these characteristics separately.

The key product attributes chosen in this case, given in Table 2.1–2, merit a brief discussion. That the product should melt snow is obvious, but a good reminder that we must soon choose a useful temperature range (−30°C to duplicate the glycol). Applying the product must be easy; spraying is one obvious alternative. The idea that the product cannot be highly toxic is self-evident, so we have chosen the more specific term "noncarcinogenic." Clearly, the product should not easily burn, but we have not made this essential because we feel that, with good practice,

TABLE 2.1–2   Characteristics of Deicing Fluids

Essential	Desirable	Useful
The product must melt snow.	The product is noncarcinogenic.	The product is inexpensive.
It must be easily applied.	It is nonflammable.	It is water miscible.
It must be noncorrosive.	It is easily recycled.	It is available from multiple suppliers.

Note: This table organizes the more scattered topics in Table 2.1–1.

we can mitigate this risk. With this table, we are poised to seek ideas for this new, recyclable deicer.

EXAMPLE 2.1–3   "SMART" LABELS

Your company currently manufactures the labels to attach to food in supermarkets. For example, the labels attached to packaged chicken could give the weight, the price per kilogram, the price of the package, and the date by which the chicken should be eaten. Although your labels are a successful product, you want to improve them so they tell the consumer more about the chicken. For example, the date on such a "smart" label might change if the chicken were frozen.

Your core team has decided that your goal is to make labels that let consumers judge product quality. The team has also decided to focus on the food and pharmaceutical areas, building on company experience with its current labels.

Identify and organize customer needs for these new smart labels.

SOLUTION

To investigate this new product, we cannot easily use our standard questions, suggested above, because we are not improving an existing product, but developing a new one. We do not have good general guidelines. Instead, we begin by assembling several peers and collecting their thoughts. Then we organize these ideas, at least in a preliminary fashion.

Our efforts to identify needs are partially summarized in Table 2.1–3. Although this list is abridged, it is not edited. In this abridgement, we did not exclude obvious tangents or irrelevancies. For example, idea 13 implies that the chemistry used for smart labels could be used for better condoms. Now it is true that there could be a major market for condoms that gave their users greater pleasure by, for example, releasing sexual stimulants. Such condoms might be more widely used, and hence reduce the spread of the HIV virus. But your current company is very unlikely to jump from food labeling into condom manufacture. This irrelevant idea has value only in waking up any who are dozing.

Once we have a collection of needs such as those in Table 2.1–3, we must organize them. In cases as unstructured as this one, it is probably best to organize them not around business topics but around intellectual topics, as shown in

TABLE 2.1–3   Customer Needs for "Smart" Labels

1.   We need labels to tell if the chicken has spoiled.
2.   Similar labels would be useful for ground beef.
3.   Smart labels should say if ice cream has ever melted.
4.   They should say if the chicken ever got warm.
5.   The best labels would be stamped onto meat, like the current USDA inspection stamps.
6.   Labels on canned goods would be good, too.
7.   Canned goods labels should detect botulism.
8.   Stick-on labels are better.
9.   Labels should remind you when drugs go bad.
10. Some labels could release drugs slowly.
11. Nicotine patches are one good smart label.
12. Could labels also release good smells?
13. This technology could be used to make better condoms.
14. Dairy goods should have some way to show their temperature history.
15. These labels could respond to lactic acid, so you could tell if the milk had spoiled.
16. You tell if things are spoiled by smelling them.
17. Labels for cream should tell you just before the cream turns sour.
18. Labels for milk should too.
19. Eggs do not spoil much.
20. Labels for frozen foods could say if they have ever been thawed.
21. Beer labels should say if it is at the right drinking temperature.
22. Wine labels could, too.
23. Labels for fish should indicate that it has no smell.
24. We need labels for mussels.
25. How can color-blind people use labels that change color?
26. Eggs should be thrown away when their stated shelf life expires.
27. Pharmaceuticals stay useful long past their stated dates.
28. Prescription drugs are usually finished, so there is no need for smart labels.
29. Could the labels be activated by the customer?
30. These would be like Post It notes that tell you when you must eat specific foods.
31. These must be activated by the consumer.
32. They could be part of tamper-proof packaging, activated automatically by opening the package.
33. Some labels could be irreversible: once they change, they should not change back.
34. The most important question is "Has this food ever been warm?"
35. These labels are especially important for caterers.
36. Labels could respond to pH changes as well as temperature changes.

Table 2.1–4. From this outline, we can target product needs consistent with our company's business interests. We could choose to develop labels that use thermally triggered, irreversible reactions as our initial focus. We could choose labels that detect spoilage or are consumer activated as secondary needs. However, choosing product needs for final development will clearly require taking these organized needs back to our customers for further definition.

TABLE 2.1–4   One Possible Organization of "Smart" Label Needs

  I. Temperature
     A. Actual temperature assessed
        Beer and other foods (21, 22)
     B. Temperature history assessed (14, 34)
        Implies irreversible reactions (33)
     C. Target: foods (3, 4, 20)
 II. Spoilage
     A. Anticipate spoilage (17, 18, 30)
     B. Detect spoilage (15, 16)
     C. Targets: foods (1, 2, 23, 24) and drugs (9, 27, 28)
III. Customer activated labels (29, 31)
 IV. Tangents
     A. Label materials (5–8)
     B. Controlled release (10–12)

Note: This table organizes the needs listed in Table 2.1–3. The numbers in parentheses refer to that table; needs 13, 19, 25, 26, and 35 are omitted.

This is a good example. As a reader, you can teach yourself much more about evaluating customer needs by repeating this example than by reading the text. At this stage of product design, you will almost certainly get results different from those given above. Indeed, when we have repeated our interviews, we also got somewhat different results. These differences disappear as we continue with the product design.

2.2   Consumer Products

Often the needs of chemical products are easily evaluated with conventional scientific instruments. We can use an ammeter to measure the current in a battery that has been shorted out to determine how long the battery separator takes to shut down. We can use a balance to measure the mass of ice melted by a particular mass of chemical. Although these measurements may require judgment, they are familiar.

In contrast, many consumer products have important characteristics that may be more difficult to measure by using conventional instruments. We may want to measure the "smoothness" of skin so that we can develop superior cosmetics. We may want to determine the "softness" of fleece during the development of new winter parkas. We will normally be less confident about these attributes, and so we may have trouble developing appropriate specifications.

In this section, we want to discuss special problems for these consumer products. We are concerned with both identifying needs and setting specifications. Identifying needs requires a method of assessing these needs. Setting specifications is facilitated by connecting consumer attributes to more familiar instrumental measurements. These problems are summarized below.

CONSUMER ASSESSMENTS

The chief problem in assessing consumer products is the scale used for evaluation. Three types of scales are useful. The first is a simple comparison test, exemplified by

"Is sample A more (attribute) than sample B?"

For example, is cream A smoother than cream B? Is meat A more tender than meat B? Is fleece A warmer than fleece B? People are good at making this type of judgment. The comparative assessments of many consumers are arithmetically averaged.

Comparison tests are especially useful for evaluating alternative products that are similar. For example, we may wish to evaluate six alternatively formulated spray starches. We can ask consumers which of a pair of starches works better. In making these comparisons, each consumer should be tested for consistency. For example, if starch A is judged better than starch B, and B is judged better than C, then A should be better than C. Such comparison tests can identify better product formulations, but they are less useful in setting specifications for new products.

A more effective scale for setting specifications assigns different products to categories. One common example of such a scale is the grading given in a course: A, B, C, and so forth. More often, the scale will use a numerical metric such as 1 to 10, where 1 is the low value and 10 is the high. Assessments of different consumers are arithmetically averaged:

$$\bar{n}_i = \sum_{j=1}^{k} n_{ij}/k, \tag{2.2-1}$$

where $\bar{n}_i$ is the average assessment of sample i made by k customers and averaged over their individual assessments n_{ij}.

This type of scale has several important characteristics that are worth remembering. First, ten different scores are as much as anyone can handle, so consumers should be discouraged from using fractional scores such as 3.1 and 8.6. Second, the average score on this scale is 5.5, not 5.0. A score of 5.0 is below average. Third, many consumers will avoid scoring anything 1 or 10, feeling that other samples that are not being considered can always be better or worse. This compresses the scale. Still, category scales are probably the most useful type.

The third common scale asks for ratios of assessments. For example, a consumer given six cookies would be asked to rate their relative sweetness. The consumer would choose one cookie as a standard and give it an arbitrary value n_0. He or she would then judge the ratio of sweetness in the other cookies relative to that of the standard. For example, cookie 1 could be twice as sweet; cookie 2 could be half as sweet; and so on. Note that assessments of zero sweetness make no sense, as all assessments are relative.

Averages of these assessments are complicated because each consumer chooses his or her own standard. To average the data for various consumers, we first calculate the values relative to the average of all assessments by one consumer:

$$n_{ij} = n_{ij}^0 \Bigg/ \left(\prod_{i=1}^{l} n_{ij}^0 \right)^{1/l}, \qquad (2.2\text{--}2)$$

where n_{ij}^0 is the actual assessment of consumer j on sample i, one of a group of l samples. We then take a geometric average over the various k consumers to get the average assessments for each sample:

$$\bar{n}_i = \left[\prod_{i=1}^{k} n_{ij} \right]^{1/k}. \qquad (2.2\text{--}3)$$

The geometric averaging results from the use of ratios, rather than categories. Although this type of scale is claimed to be superior for fundamental studies of perception, it is less common than the category scale in product development. Most use category scales.

Once we have chosen a scale, we will present consumers with a selection of possible products and ask for their evaluations in term of popular vocabulary. The vocabulary depends on the specific products. For example, breakfast cereal is described as "crunchy," wool substitutes can be judged as "soft," and skin creams aim to make skin "smooth." We then average the consumer responses. Alternative methods of averaging and ways to screen anomalies in the data are beyond the scope of this book, but they are extensively discussed in the marketing and psychophysical literature. With these average results, we can now formulate a ranking of consumer needs.

CONSUMER VERSUS INSTRUMENTAL ASSESSMENTS

We next want to reformulate these consumer needs as more quantitative specifications. The qualitative needs may well be trite: we already know that consumers want crunchier cereal, softer woollens, and smoother skin. The question is how much improvement will be noticeable and valued.

Making the jump from "better" to "how much better" is normally easier if we can relate specific attributes to particular scientific parameters. To be sure, we can continue to guess new product formulations, and then run back to our consumers to ask which formulation they like best. Such repeated consumer evaluation can work, but it is slow and expensive. In product development, where reducing development time is a constant goal, such a strategy should be avoided. Instruments are better.

The connection between consumer attributes and instrumental measurements is normally sought empirically. Often, this connection is exactly what we would expect. For example, "thick" soups have high viscosity and "thin" soups have low viscosity. This implies that assessments of "thick" should correlate inversely with assessments of "thin" but directly with viscosity measurements. This is supported by experiment. In the same way, assessments of "sweetness" correlate with measured sugar concentrations and judgments of "sourness" are proportional to citric acid concentration.

Sometimes, the connection between consumer attributes and instrumental measurements is unexpected. For example, we might expect that a breakfast cereal's "crunchiness" would be fairly directly related to the cereal flakes' fracture mechanics. It is not. Instead, it correlates best with the sound released during chewing. In the same way, we might expect that a beer's "smoothness" would be related to its viscosity. It is not. It is associated with the force of contact lubrication. This is why beers sold as "smooth" often have many very small bubbles. These small bubbles reduce the coefficient of friction on the tongue and hence the related contact lubrication force.

The uneven correlation between consumer assessments and instrumental measurements has spawned an enormous number of proprietary instruments, which try to imitate consumer assessments more exactly. Some of these are bizarre. One famous example attached a pair of dentures to a rheometer. To use the rheometer, one would place a sample of, for example, beefsteak between the dentures and measure the force versus distance. The result would be a measure of the beef's rheology and hopefully of attributes like its tenderness.

Occasionally, we can see how consumer assessments and the instrumental measurements are quantitatively related. For example, imagine that you pick up a metal spoon and a wooden spoon from a kitchen drawer. The metal spoon feels colder than the wooden one. In fact, both spoons are at the same temperature – the room temperature. Thus your perception is not of the spoons' temperature, but of your skin's temperature. If you estimate how your skin's temperature is influenced by the thermal diffusivity of the spoon, you can predict how much colder the metal spoon feels compared with the wooden one. Two further examples of relations between consumer attributes and instrumental measurements close this section.

EXAMPLE 2.2–1   TASTY CHOCOLATE

One of the attractive factors of good chocolate is its smooth "melt-in-the-mouth" quality. Some chocaholics even claim a pleasant cooling of the mouth, as the fat crystals in the chocolate absorb latent heat and melt. Sometimes chocolate exhibits a powder tan layer called "blooming," especially when it is old and has been stored at fluctuating temperature. This chocolate has identical ingredients and chemical composition to tasty chocolate, but it has an unpleasant powdery texture in the mouth.

Chocolate manufacturers, such as Nestlé, employ expert panels in order to assess the quality of their chocolate. These testers are extremely competent – some can even identify the country of origin of the cocoa beans, in much the same way an expert oenologist can tell the chateau of a fine wine. The testers do a good job. However, it would be quicker, cheaper, and perhaps more reliable if we could augment some of their assessments by using scientific instruments.

SOLUTION

Chocolate gets its melt-in-the-mouth sensation from the melting of cocoa butter crystals. Cocoa butter is a triglyceride, similar to olive oil, margarine, and animal

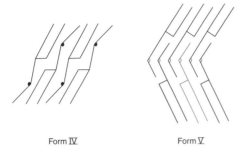

Figure 2.2–1. Schematic Structures of Forms IV and V of Cocoa Butter. Individual triglyceride molecules adopt "chair" structures. In Form IV, these chairs stack with a repeat distance approximately two times the length of a single fatty acid chain. In Form V, this repeat distance is about three times that in Form IV. Form V tastes better.

fat. Although triglycerides are the commonest form of naturally occurring fat, most have a poorly defined melting point because they contain a range of different fatty acids. Cocoa butter is unusual in having a narrow range of composition; it mainly consists of symmetrical triglycerides with oleic acid (unsaturated) in the 2- position and palmitic or stearic acid (saturated) in the 1- and 3- positions. This results in a well-defined crystal structure and a sharp melting point. However, cocoa butter can crystallize into five different crystal forms, each with its own melting point. Which one is formed depends on the exact preparation recipe. This is why confectionery chefs worry so much about cooling rate and stirring speed when preparing chocolate dishes from the melt. Form V is the desirable one, with a melting point of about 35°C, just below mouth temperature. Form IV, produced by cooling rapidly without stirring, has a melting point of about 28°C and is believed to be the cause of blooming. The structures of forms IV and V are shown schematically in Figure 2.2–1.

In effect, chocolate testers spend a lot of effort in identifying crystal forms of cocoa butter. We can do this more reliably and quicker with the use of differential scanning calorimetry (DSC). In this method we measure the rate of heat flow as the temperature is raised: as the crystals melt, energy is absorbed in the form of latent heat, without a concomitant temperature increase. We identify the presence of different crystal forms by the presence of different melting points in DSC. If we wish to be more quantitative, we can use the more sophisticated and expensive technique of powder X-ray diffraction, which shows how much of each crystal form is present. We can then assess product quality and so optimize manufacturing and storage procedures.

EXAMPLE 2.2–2   THE CONSUMER ATTRIBUTE "VISCOSITY"

Studies of skin creams show that the consumer attributes of "thick" and "viscous" are closely related and that both are correlated with instrumental measurements of viscosity. However, the correlation is not linear:

$$\text{(assessments of "thick")} \propto \text{(assessments of "viscous")}$$
$$\propto \sqrt{\text{measurements of viscosity.}}$$

Suggest why this could occur.

SOLUTION

The obvious hypothesis is that "viscous" skin creams cause larger forces when rubbed on the skin. The force F is related to the viscosity μ by

$$\text{(assessments of "viscous")} \propto F = -\mu A \frac{dv}{dy},$$

where A is the area of the fingers, v is the fluid velocity, and y is the distance normal to the finger surface. The area is constant; because the layer of skin cream is thin, the derivative equals the finger speed v divided by the cream thickness h. Thus

$$\text{(assessments of "viscous")} \propto \frac{\mu v}{h}.$$

However, more viscous creams will have a thicker layer h on the surface. In fact, from a calculation by Clerk Maxwell, we may show that h is proportional to $\mu^{1/2}$. Thus, because the finger velocity v is observed to be constant,

$$\text{(assessments of "viscous")} \propto \frac{\mu v}{\mu^{1/2}} \propto \mu^{1/2}.$$

Similar arguments for power-law fluids are consistent with experiments, lending further support to this hypothesis.

2.3   Converting Needs to Specifications

The customer needs that we discover during interviews are usually qualitative. These needs may include trivial product changes and unrealistic product dreams. Their assembly and editing will be dominated by those whose strengths lie in marketing. Those on the core team whose strengths are in chemistry and engineering can be good critics or cheerleaders, but they will depend on others for detailed expertise.

Our next step is to convert these qualitative needs into particular product specifications. For example, if we want to make the airplane deicers, we must decide on the degree of freezing point depression, the speed of melting, and the efficiency of any recycle. Marketing is now less important and chemistry and engineering are paramount.

We can benefit from a cogent strategy for setting specifications. The experienced designer may use this strategy merely as a checklist for his or her own creativity. The novice may use it as a convenient way just to get started. The strategy we suggest has three steps:

1. Write complete chemical reactions for any chemical steps involved.
2. Make mass and energy balances important in product use.
3. Estimate any important rates that occur during product use.

Many of you with technical training will recognize that this strategy is a précis of the undergraduate curriculum.

Our rationale for these three steps merits discussion. Carefully writing out the equations of any chemical reactions may seem silly, especially because the reactions in some chemical mixtures may be incompletely defined. Many chemists will find the exercise trivial. Nonetheless, in our experience, this step is important to force consideration of mass versus moles. We have found that those without chemical training often write specifications in terms of mass, though the actual chemical change is in moles. These confusions are not always limited to nonchemists. We vividly remember one petrochemical research group carefully developing a partial oxidation catalyst based on a hemoglobin analog, only to realize that the analog's large molar volume meant that it would have 1000 times too few active sites per volume to be practical. The first strategic step should always be checking the stoichiometry.

The second step is identifying the important mass and energy balances in the system. Such a thermodynamically grounded step establishes what can happen. It will give the maximum volume change, or the final temperature after an adiabatic reaction, or the size of any process recycles. In making these balances, we seek the simplest nontrivial solution. All physical properties, such as densities and heat capacities, are taken as constant. All solutions should be assumed to be ideal. Like the first step, this will not take long.

The third step, the rate processes, should again make as many simplifying assumptions as possible. All relevant transport coefficients, such as the thermal conductivity and the diffusion coefficient, should be taken as constants. The viscosity should be assumed Newtonian unless non-Newtonian behavior is essential to the specifications. Chemical reactions should be taken as first order or zero order. Reaction rates should be capped: they cannot go faster than diffusion control, even at high temperatures. At this stage, we want a simple answer, as exemplified by the following examples.

EXAMPLE 2.3–1   MUFFLER DESIGN

Automobile mufflers (in Britain, "silencers") rust from the inside out. They do this because after the car is driven, the muffler contains exhaust gases, including water vapor. When the muffler cools, the water condenses and corrodes the inside of the muffler.

One clever route to avoid this problem is to put a small bag containing hydrophilic zeolite in the muffler, as shown in Figure 2.3–1. This adsorbs the water vapor, preventing liquid condensation and so dramatically reducing corrosion.

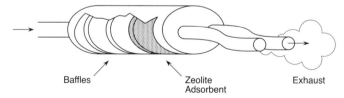

Baffles   Zeolite   Exhaust
                 Adsorbent

Figure 2.3–1. Car Mufflers with Reduced Corrosion. The proposed muffler would contain a small amount of zeolite, which absorbs water left in the muffler when the car's engine is shut off. The water is released when the car is restarted and the muffler again gets hot.

When the car is restarted, the hot exhaust gases heat the zeolite and drive off the adsorbed water. The zeolite is then ready to adsorb more water when the engine is stopped.

We are considering making a muffler that has this feature. How much water will we need to adsorb? How fast should the adsorption be?

SOLUTION

This problem is simply an exercise in stoichiometry. Assume the muffler's volume is about 5 L, and that is it about 70% voids. The basic reaction in the engine is

$$C_8H_{18} + \frac{25}{2}O_2 \rightarrow 8CO_2 + 9H_2O.$$

If the engine is run with 10% excess air, then the exhaust concentration at complete combustion is easily shown to be 0% C_8H_{18}, 2% O_2, 75% N_2, 11% CO_2, and 12% H_2O. Thus

$$\left(\frac{5\,L}{22.4\,L/gmol}\right)0.7\left(\frac{0.2\,gmol\,H_2O}{gmol\,total}\right)\frac{18\,g}{gmol} = 0.34\,g\,H_2O.$$

Adsorbing this small amount can prolong the life of the muffler. We want the adsorption to be prompt, but it need not be faster than the time that the muffler takes to cool. Thirty seconds is probably a reasonable starting point.

EXAMPLE 2.3–2 WATER PURIFICATION FOR THE TRAVELER

People traveling into wilderness areas require drinking water. Often water sources such as streams and ponds are contaminated by viruses and bacteria. A particular problem in North America is giardia, a protozoan present in 90% of water sources. (It is carried by animals as well as humans, which explains its wide spread.) Giardia results in severe intestinal problems, with very unpleasant symptoms including noxious farts with a sulfurous tang, which is particularly unfortunate if you are sharing a tent.

Interviews with customers (hikers, mountaineers, soldiers, and equipment suppliers) might reveal the following list of needs in a water purification device:

Produces safe water
Is light and small
Is fast acting
Has a long lifetime
Requires no power source
Is cheap and reusable
Improves odor and flavor

Write specifications for a water purification product.

SOLUTION

This problem is somewhat different from that just discussed because we are now considering a consumer product. For such products, the two steps of chemical reactions and thermodynamics suggested above are not helpful. The chemical reactions and mass and energy balances are trivial. There is a rate process at work: how fast do people lose water? The answer is about 5 L or 1 gal/day. Indeed, the original definition of a gallon derived from the daily allowance of water a farmer had to provide to each laborer.

This starts off our quantification, which in this case will be led more by intuition than science. We might aim to design for groups of two to four persons, for trips of up to 2 months. If each person needs around 5 L of drinking water per day, we need to be able to purify 2000 L of water before our product fails. We need around 20 L per day, but probably we would like to produce it all quickly, for example for cooking, so we want a flow rate of say 1 L min^{-1}. Because the product must be carried up mountains, it must weight less than 1 kg and occupy less than 1 L volume.

For safety we can look to legislation: U.S. health regulations require 99.9% removal of bacteria and protozoa in surface water treatment. Note we have focused here on bacteria and protozoa. This is appropriate in the North American wilderness, but if we were considering travelers in third world countries, we also need to worry about waterborne viruses, passed on through human waste. Here we have an example of different specifications for different markets. Price? Climbers will pay in excess of $100 for a good ice axe, so one might expect they would be willing to fork out this much for clean water. The product must be effective over a full range of temperatures at which water is found and at altitudes up to 7000 m: 0–40°C and one third of 1 atm pressure.

Our final specifications might take the following form:

Has a capacity of 2000 L
Has a production rate of 1 L min^{-1}
Removes 99.9% of bacteria and protozoa
Costs less than $100
Has an operating range of 0–40°C, one third of 1 atm
Improves odor and flavor

We are now in a good position to start thinking of ways to achieve these specifications.

EXAMPLE 2.3–3 PREVENTING EXPLOSIONS IN HIGH-PERFORMANCE BATTERIES

Modern electronic devices such as cellular phones and laptop computers require powerful batteries in order to function without frequent recharging. The result, shown schematically in Figure 2.3–2, is batteries with ever increasing power per volume and power per mass. Putting more and more energy into smaller and smaller packages means that the batteries are becoming bombs. Indeed, the president of one battery manufacturer had his car phone blow up in his ear while he was talking to his wife.

Obviously, there is a need for safe batteries that shut down before they blow up. How fast does the battery need to shut down?

a)
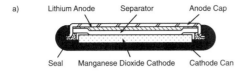

b)
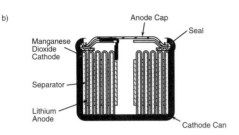

Figure 2.3–2. High-Performance Batteries for Laptop Computers. The high energy density of these batteries, especially those for laptop computers, means that accidental short circuits can cause explosions. The coin cell in a) is drawn on a larger scale than the cylindrical cell in b).

SOLUTION

In answering this question, we consider the most powerful batteries available, which are based on lithium. Because chemical details of lithium batteries are hard to come by, we choose alkaline nickel-cadmium (NiCad) batteries as our standard. We can then extrapolate our results to lithium batteries, which are about five times more powerful. We also choose a laptop computer like the IBM Thinkpad as a standard application. This computer currently uses a battery with about 50 Whr energy.

We begin by writing out the basic reactions during discharge. At the anode:

$$Cd + 2OH^- \rightarrow Cd(OH)_2 + 2e^-.$$

At the cathode:

$$Ni(OOH) + H_2O + e^- \rightarrow Ni(OH)_2 + OH^-.$$

The overall reaction is thus

$$2Ni(OOH) + Cd + 2H_2O \rightarrow 2Ni(OH)_2 + Cd(OH)_2.$$

This reaction is reversed during charging. The electrodes are separated by aqueous KOH, with a density of around 1.2 g/cm^3. This roughly corresponds to a concentration of 4 M. Under normal operations, the hydroxide concentration is nearly constant between the electrodes.

We now turn to the mass and energy balances. A battery in a laptop typically contains two electrodes around 3 mm thick, separated by a membrane separator. The membrane separator, which is around 30 μm thick, contains about 30% pores filled with the 4 M KOH. The battery is usually about 20 cm square, so its total volume is around 240 cm^3.

The adiabatic temperature rise if the battery is shorted out is found from the first law of thermodynamics:

$$\Delta U = Q + W,$$

where ΔU is the internal energy change, Q is the heat, and W is the work. Because the battery is adiabatic, Q is zero. Because the battery has a constant volume, W is zero. Because the battery contains no gas, ΔU is about equal to the enthalpy change ΔH. Thus

$$\Delta U = 0 = \Delta H_{rxn} + V\rho\hat{C}_p\Delta T,$$

where ΔH_{rxn} is the heat of reaction under the initial conditions, V is the battery volume, ρ is its average density, $\hat{C}_p$ is its average specific heat capacity, and ΔT is the temperature rise caused by the short. Consistent with our goal of keeping our analysis simple, we assume that the battery is initially at 25°C, the normal reference temperature for chemical reactions.

The heat of reaction is the energy in the battery or $(-50\ \text{Whr})$. An average value of $\rho\hat{C}_p$ is about 3 J/cm^3 °C. Thus

$$0 = -50\frac{\text{J hr}}{\text{sec}}\left(\frac{3600\ \text{sec}}{\text{hr}}\right) + 240\ \text{cm}^3\left(\frac{3\ \text{J}}{\text{cm}^3\ °\text{C}}\right)(\Delta T)$$

and

$$\Delta T = 250°\text{C}.$$

At this temperature, the water will have turned to steam and the battery will explode. We may be in trouble. This completes our summary of the mass and energy balances.

Next, we turn to calculating the rate at which the battery heats up when it is shorted out. When the battery is shorted, the electrochemical reactions shown above are controlled by the rate of hydroxide diffusion from one electrode to the other. Thus the time taken to discharge a shorted battery is roughly given by

$$(\text{amount Cd in battery})\left(\frac{2\ \text{mol OH}^-}{\text{mol Cd}}\right) = \left(\frac{\text{OH}^-\ \text{flux}}{\text{area time}}\right)(\text{area})\text{time} = j_1 At$$

$$= \left(\frac{D}{\delta}\Delta c\right)At,$$

where j_1 is the hydroxide flux, D is its diffusion coefficient, δ is the thickness between the electrodes, Δc is the hydroxide concentration difference, and A is the battery area. NiCad batteries normally have about 100 C Cd/cm^3 in the anode, which occupies roughly half the battery volume; the diffusion coefficient of the hydroxide is about 5×10^{-5} cm^2/sec. Thus

$$\frac{100\ \text{C Cd}}{\text{cm}^3}\left(\frac{1/2(240\ \text{cm}^3)}{96500\ \text{C/mol}}\right)\left(\frac{2\ \text{mol OH}^-}{\text{mol Cd}}\right)$$

$$= \left(\frac{5 \times 10^{-5}\ \text{cm}^2/\text{sec}}{30 \times 10^{-4}\ \text{cm}}\frac{4\ \text{mol OH}^-}{10^3\ \text{cm}^3}\right)400\ \text{cm}^2\ t;$$

$$t = 10\ \text{sec}.$$

However, we must shut down the battery before any vaporization. Because we start at 25°C, we can only stand a temperature change of 75°C. Thus

$$t = \frac{75}{250}(10) = 3\ \text{sec}.$$

The customer's need is not to be blown up. The resulting product specification is to shut down the battery within about 3 sec.

2.4   Revising Product Specifications

The strategy above – stoichiometry, overall balances, and rates – leads to preliminary product specifications. We will normally be able to obtain good estimates of these. However, such estimates often have two serious shortcomings. The first shortcoming is that our initial specifications may be blatantly unrealistic. They may suggest that we need materials which are excessively expensive. They may require huge flows or huge concentrations. They may imply elements which are not in the periodic table. Such shortcomings mean that we must revise the product specifications (or abandon the product development).

Parenthetically, this type of revision is described idiomatically with different images in different companies. Sometimes, it is called a "sanity check" or a "gut check." Often it is described as a "back of the envelope calculation." One company calls this revision a "chicken test," an apparent reference to a fabled test of aircraft engines made by the Canadian government. The test involved tossing frozen chickens into the running jet turbines. The British navy has a similar test to check out the viability of new marine impeller designs; they throw a railway tie into the propeller and reject those designs that fail to survive intact.

The second shortcoming of the strategy in the previous section is that it does not make a careful comparison with existing products. Such an existing product is often called a "benchmark." In many cases, we will have a good idea about what particular improvements we want, and we will have several specific ways in which we want to make these improvements. We can gauge the significance of these improvements by comparing our specifications with the benchmark. The comparison may show that our new specifications are inadequate or overly ambitious. If so, we should revise them.

The use of a benchmark is vividly illustrated in the first example given below, which seeks a deicer for winter roads that causes less corrosion. (We know, we had another deicing problem in an earlier section; but this book was partially written in Minnesota.) Roads are currently deiced with rock salt. Thus rock salt is an obvious benchmark, and one that is an enormous help in revising our specifications.

In some cases, we may be developing a new product for which no direct competition exists. In this case, we will have no obvious choice of a benchmark, and hence no clear guide for our specifications. We can sometimes benefit by looking for products that perform similar but chemically different functions.

The second example in this section shows how this "similar function" strategy works. We seek a liquid that selectively absorbs nitrogen from natural gas. We currently have no such liquid. However, other liquids are available that do absorb other chemical species from other gas mixtures. Thus, we choose as a benchmark aqueous solutions of monoethanol amine, which is used to absorb carbon dioxide from gas mixtures found in petrochemical processing. Comparing our hypothetical nitrogen absorber with monoethanol amine helps us to revise specifications.

We should recognize that in choosing benchmarks, we are taking a serious risk. The risk is that we will skip the product development process and jump straight to our best guess of a solution. We will not finish carefully setting specifications, searching for new ideas, selecting rationally among these ideas, and developing our chosen product. This risk is real. To minimize it, we urge the product development teams to downplay their favorite guesses when they seek product ideas in the next stage. At the same time, we urge revising specifications with any extra information that seems important to the team.

The strategy for revising the specifications recapitulates that for setting the original specifications: write out the chemical reactions, make the relevant mass and energy balances, and estimate any key rate processes. It is illustrated in the examples that follow. These examples are necessarily more complex, because they include both setting the original specifications and then revising them.

EXAMPLE 2.4–1   DEICING WINTER ROADS

In winter, roads in areas of frequent frost are spread with a mixture of sand and salt to improve traction and melt ice and snow. Although this treatment works well to about $-20°C$, the salt causes significant environmental damage. It corrodes cars about four times faster than water alone. It weakens bridge decks and parking ramps, sometimes causing their collapse. It can pollute local water wells.

Not surprisingly, government agencies frequently look for alternative chemicals to salt that are environmentally less abusive. Using salt as a benchmark, they suggest the following specifications for the alternative chemical:

1.   It should melt ice over a similar temperature range.
2.   It should melt a comparable amount of ice per kilogram as salt does.
3.   It should cause less corrosion per kilogram than salt.

Two alternatives to salt that are frequently suggested are urea and calcium magnesium acetate, often called "CMA."

Compare the performance of salt with these alternatives as a means of revising product specifications.

SOLUTION

To make this comparison, we must first consider why chemicals cause ice to melt. The basic phenomenon is freezing point depression. When any solute is dissolved in water, the chemical potential of the water molecules is lowered. As the temperature is lowered, this chemical potential also drops, but not as fast as that of ice does. When the chemical potential of the water in solution becomes higher that that of ice, the water freezes.

To put these ideas on a more quantitative basis, we remember that at the freezing temperature T, the chemical potential μ_1 of the water in the ice equals that in a solution of water of mole fraction x_1,

$$\mu_1(\text{ice, } T) = \mu_1(\text{pure liquid water, } T) + RT \ln x_1.$$

This implies an ideal solution, consistent with our strategy of choosing the simplest nontrivial case. Because the solution is dilute, x_1 is close to unity and

$$ln\, x_1 = ln\,(1 - x_2) \doteq -x_2 - \cdots,$$

where x_2 is the mole fraction of the chemical. Combining these expressions gives

$$x_2 = \frac{\mu_1(\text{pure liquid water}, T) - \mu_1(\text{ice}, T)}{RT}.$$

We must now calculate the chemical potential difference.

The calculation of this chemical potential difference is easily reviewed, but those who want to avoid the review can skip this paragraph. By definition,

$$\frac{\mu_1}{RT} = \frac{\bar{H}_1 - T\bar{S}_1}{RT} = \frac{\bar{H}_1}{RT} - \frac{\bar{S}_1}{R}.$$

In the case of a pure substance,

$$\mu_1 = \underline{G}_1, \quad \bar{H}_1 = \underline{H}_1, \quad \bar{S}_1 = \underline{S}_1,$$

where $\bar{H}_1$ and $\bar{S}_1$ are the partial molar enthalpy and entropy, respectively; and $\underline{G}_1$, $\underline{H}_1$, and $\underline{S}_1$ are the molar Gibbs energy, enthalpy, and entropy of pure 1. As a result,

$$\frac{\mu_1(\text{pure liquid water}, T) - \mu_1(\text{ice}, T)}{RT}$$
$$= \frac{\underline{H}_1(\text{pure liquid water}, T) - \underline{H}_1(\text{ice}, T)}{RT}$$
$$- \frac{\underline{S}_1(\text{pure liquid water}, T) - \underline{S}_1(\text{ice}, T)}{R}.$$

At the normal freezing point T_0,

$$\frac{\mu_1^0(\text{pure liquid water}, T_0) - \mu_1^0(\text{ice}, T_0)}{RT_0} = 0,$$
$$= \frac{\underline{H}_1(\text{pure liquid water}, T_0) - \underline{H}_1(\text{ice}, T_0)}{RT_0}$$
$$- \frac{\underline{S}_1(\text{pure liquid water}, T_0) - \underline{S}_1(\text{ice}, T_0)}{R}.$$

Subtracting these equations, and assuming the enthalpy and entropy of fusion to be temperature independent, we find

$$\frac{\mu_1(\text{pure liquid water}, T) - \mu_1(\text{ice}, T)}{RT}$$
$$= \{\underline{H}_1(\text{pure liquid water}, T) - \underline{H}_1(\text{ice}, T)\}\left(\frac{1}{RT} - \frac{1}{RT_0}\right)$$
$$= \{\Delta\underline{H}_{\text{fusion}}\}\left(\frac{1}{RT} - \frac{1}{RT_0}\right)$$
$$\doteq \left\{\frac{\Delta\underline{H}_{\text{fusion}}}{RT^2}\right\}(T_0 - T)$$

In this last equality, we have recognized that the freezing point depression is much smaller than the absolute temperature.

We now combine this free energy difference with our previous result to find

$$x_2 = \frac{\Delta H_{\text{fusion}}}{RT^2}(T_0 - T).$$

This result, which may also be derived from the Gibbs-Helmholtz equation, shows how the solution concentration is related to the freezing point depression. For example, for a drop of $10°K$,

$$x_2 = \frac{6 \text{ kJ/mol}}{8.31(\text{J/mol}°\text{K})(273°\text{K})^2}(10°\text{K})$$
$$= 0.097.$$

For salt, this is about equal to 18 g of NaCl per 100 g of H_2O. This result is useful for revising the first two product specifications given above.

Developing a basis that is similarly useful for the third, corrosion-based specification is more difficult. The corrosion rate will normally involve two or more sequential chemical steps. In the limit where surface chemical reactions are controlling, the rate of corrosion will be given by the Butler-Volmer equation, which for small overpotential η is

$$j_1 = j_0 \eta \mathcal{F}/RT,$$

where j_0 is the exchange current density, that is, the reaction rate at equilibrium, and $\mathcal{F}$ is Faraday's constant. Note that the corrosion rate is a flux, with units of moles per area per time. Unfortunately, we have no easy ways of guessing either j_0 or η in this situation, so this result does not help us in our revision.

We can estimate the fastest possible corrosion rate, which will occur when paint or rust limits how fast the deicing chemical can reach the surface. This rate is the diffusion controlled limit:

$$j_1 = \left(\frac{D}{\delta}\right)c_1,$$

where D is the diffusion coefficient of the chemical across a rust layer of thickness δ. Note that c_1 is the molar concentration, not the mass concentration. Similarly, in the freezing point depression, the important variable was the molar concentration.

We can now compare the expected performance of the three possible chemicals. Although salt and urea are straightforward, CMA requires some explanation. This material, which is not readily available commercially, is made by reacting dolomite limestone and acetic acid:

$$CaMgCO_3 + 4CH_3COOH \rightarrow (CH_3COO)_4CaMg + 2CO_2 + 2H_2O.$$

The limestone sells for around $20/ton; the acetic acid is about $0.37/lb; so the material cost is almost $800/ton. Clearly, this cost is dominated by the acetic acid and can be reduced if waste acetic acid is available.

The results of the comparison of salt, urea, and CMA are given in Table 2.4–1. Right away, we see that salt will be difficult to beat if the initial specifications are

TABLE 2.4–1   Relevant Properties of Three Possible Deicers

Basic Compound	NaCl	Urea	CMA
Solubility g/100 g H_2O	36	100	40
Species in solution	Na^+, Cl^-	$O=C\!\!\begin{smallmatrix}NH_2\\NH_2\end{smallmatrix}$	Ca^{++}, Mg^{++} $4CH_3COO^-$
Molecular weight	58.5	60	300
Moles particles/mole compound	2	1	6
Mole fraction at saturation	0.18	0.23	0.13
Moles particles/mass compound	34	17	20
Cost per metric ton, U.S. dollars	20	130	800
Moles particles per dollar	1700	130	25

Note: Sodium chloride is the benchmark. Urea is sometimes used on airport runways because of reduced corrosion. CMA (($CH_3COO)_4$ CaMg) is one alternative favored by some government agencies.

used. In particular, the first specification says that an alternative chemical should melt ice over a similar temperature range to salt. From the equations above, we see that the maximum freezing point depression is proportional to the maximum mole fraction, that is, the mole fraction of saturation. From Table 2.4–1, we see that these mole fractions are roughly comparable, so we seem to be on the right track.

However, the second specification is that a fixed mass of chemical should melt the same amount as salt. Here, the alternatives stumble. Although urea has a similar molecular weight as salt, it does not ionize, so a mass of urea gives a solution whose mole fraction is about half that of salt. In the same way, CMA produces only half the mole fraction as salt, or in the terms used in the table, 20 versus 34 particles per mass. Salt is better.

The third specification, regarding corrosion, is the hardest to evaluate. The corrosion from urea will be modest, because urea does not produce ions. Salt and CMA do produce ions. When corrosion is diffusion controlled, the corrosion rate is proportional to the ionic concentration. But ice melting is also proportional to ionic concentration. Thus any ionic chemical will accelerate corrosion and melt ice in roughly proportional amounts.

Finally, we turn to chemical cost. Although cost was not included in our original specifications, it will certainly have a role in our final decision. What will matter is the ice melted per dollar spent, or in chemical terms, the particles delivered per dollar spent. As Table 2.4–1 shows, salt is way out in front.

Thus if we still seek an alternative chemical, our specification requires significant revision. The exact nature of this revision depends on getting additional information. Even without this, we can see alternatives requiring creative thinking. For example, we are immediately struck by the high cost of the alternative chemicals. Urea is already a commodity, so we are unlikely to be able to cut its cost much. On the other hand, CMA's price is dominated by the acetic acid cost.

To cut this cost, one group creatively fermented a mixture of lime and garbage in cement mixers that normally sat unused in the winter. They then used the fermented garbage as a deicer, a solution better suited for rural areas.

As a second example, we can remember that our real objective is to clear the road, not necessarily by melting the ice. If we can just get the ice to debond from the road surface, then we can remove it with a snowplow. This suggests replacing the specification of ice melted per mass of chemical with ice removed per mass in some standard test. Such a revised specification makes good sense.

EXAMPLE 2.4–2   SCRUBBING NITROGEN FROM NATURAL GAS

In the next few decades, a major fraction of the world's energy is expected to come from natural gas. The best natural gas is largely methane, with small amounts of other hydrocarbons. Future natural gas may be much less pure, containing large amounts of carbon dioxide and nitrogen.

These impurities have to be removed. One easy way is cryogenic distillation, but this is prohibitively expensive. Carbon dioxide can be removed by absorption into aqueous solutions of amines, using packed towers like that in Figure 2.4–1. There is no similar method of absorption for nitrogen.

Thus the need is for a liquid or liquid solution that absorbs nitrogen but not methane. There is currently no liquid with this property, so there is no obvious benchmark. Instead, we choose monoethanol amine, the standard absorbent for carbon dioxide. Our first attempt at product specification seeks a new liquid solution for nitrogen that would be used as the amine solutions are used for carbon dioxide. This new solution would have similar loading, react with similar kinetics, and use similar equipment to that used for CO_2. Such similarity is sometimes called a "transparent technology" because the operator running the process does not need to know if he or she is removing CO_2 or N_2.

See if these specifications are reasonable, and revise them if necessary.

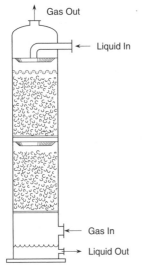

Figure 2.4–1. Packed Towers for Carbon Dioxide Adsorption. The gas containing carbon dioxide flows upward, and an aqueous solution of reactive amines flows countercurrently downward. The reaction facilitates the removal of the carbon dioxide.

SOLUTION

We can begin by looking at the stoichiometry and the mass balances for carbon dioxide. The key reaction is

$$CO_2 + H_2O + 2R\,NH_2 \rightarrow (R\,NH_3)_2CO_3,$$

where R is, for example, $HOCH_2CH_2$. If the aqueous solution contains 10 wt.% amine and reacts completely, then the amount of CO_2 absorbed will be

$$(10\%)\left(\frac{1\ \text{mol}\ CO_2}{2\ \text{mol amine}}\right)\left[\frac{44\ \text{g}\ CO_2/\text{mol}}{61\ \text{g amine/mol}}\right] = 3.6\%\ CO_2.$$

To meet out specifications, we would like our new liquid to absorb a similar amount of nitrogen.

To see if this specification is reasonable, we need to anticipate the chemistry of nitrogen complex formation. The complexes are most likely going to be organo-metallic, possibly with a porphyrin structure aping that in hemoglobin. These chemical compounds are large, with a molecular weight around 500 or more. They are very unlikely to be soluble in water to greater than 10 wt. %. Thus, if the organometallic compound forms a 1:1 complex,

$$10\%\left(\frac{1\ \text{mol}\ N_2}{1\ \text{mol complex}}\right)\left[\frac{28\ \text{g}\ N_2/\text{mol}}{500\ \text{g complex/mol}}\right] = 0.6\%\ N_2.$$

Like it or not, the nitrogen scrubbing solution is going to be more dilute.

Justification for the 10 wt. % figure as the likely limit we will be able to achieve for the solubility of our nitrogen absorbing complex comes from animals' oxygen absorbing systems. Animals without corpuscles achieve pigment concentrations in their blood of up to about 5 wt.%. Red corpuscles allow pigment concentrations in the blood stream to be raised above this figure (it is about 15 wt. % in human blood), making high levels of aerobic activity sustainable. See Figure 2.4–2 for examples of the oxygen absorbing capacity of animals. Squid have no blood corpuscles; they are therefore vulnerable to predation by fish, which are equipped with corpuscles. To compensate, squid have three hearts in order to speed up their circulation, and adaptations such as ink release to confuse predators. The evo-lutionary pressure to maximize the oxygen carrying capacity of blood has been strong. We are most unlikely to be able to develop a heme analog with a higher nitrogen carrying capacity; if anything, 10 wt. % represents an optimistic value.

The relative dilution of our absorber, a consequence of reaction stoichiometry, has a significant effect on the rates of absorption as well. There are two of types of rate processes that are important. First, the new complex must react with the nitrogen quickly and reversibly. Ideally, we would like complex formation to be as fast or faster than the amine-CO_2 reaction. Equally, we would like to have the complex quickly decompose when we increase the liquid temperature slightly. To discover if this can happen, we must synthesize our complexing agent and make absorption and desorption experiments. The results of these experiments will determine the height of the packed towers that we will require for absorption and stripping.

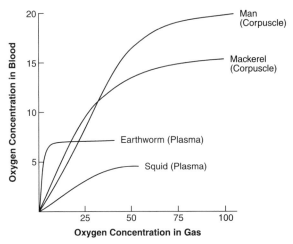

Figure 2.4–2. The Oxygen Absorbing Capacity of Various Animals' Blood. Animals without red blood corpuscles, like the squid, achieve substantially lower oxygen loading at saturation than those with corpuscles, like the mackerel. Units are mol O_2 in 100 mol of blood and mm Hg partial pressure of O_2.

The second type of rate process concerns not the height but the diameter of the packed tower. To explore this, we return to our carbon dioxide benchmark. Imagine that we have a gas flux G of density ρ_G and a liquid flux L of density ρ_L flowing in a packed tower of cross section A. As a basis, we assume that we have a total gas flow equal to 100 mol/min. This feed stream contains 80 vol. % CH_4 and 20 vol. % CO_2. Such a feed will have an average molecular weight of 21.6 g/mol. If 90% of the CO_2 is removed to produce an amine solution which is 50% saturated, then the mass flow of amine solution LA may be found from

$$(\text{mass } CO_2 \text{ leaving in liquid}) = (\text{mass } CO_2 \text{ absorbed from gas}),$$

$$0.50(0.036)LA = 0.90(0.20)\frac{100 \text{ mol}}{\text{min}}\left(\frac{44 \times 10^{-3} \text{ kg}}{\text{mol}}\right),$$

$$LA = 44 \text{ kg/min}.$$

The operation of the packed tower depends critically on the "flow parameter," which for this feed is

$$\frac{LA}{GA}\sqrt{\frac{\rho_G}{\rho_L}} = \frac{44 \text{ kg/min}}{(100 \text{ mol/min})[(0.0216 \text{ kg/mol})]}\sqrt{\frac{21.6 \text{ g}/22.4 \times 10^3 \text{ cm}^3}{1 \text{ g/cm}^3}},$$

$$= 0.63.$$

This value, in the middle of the normal design range, can be used to determine the tower's cross-sectional area.

We now turn from the benchmark, carbon dioxide, to our new target, nitrogen. As before, we imagine that we have the 100 mol/min feed, which is now 80% CH_4 and 20% N_2, and 90% of the N_2 is removed to produce a liquid that is again 50%

saturated. The mass flow of this new product solution is

$$0.50(0.006)LA = 0.90(0.20)\frac{100 \text{ mol}}{\text{min}}\left(\frac{28 \times 10^{-3} \text{ kg}}{\text{mol}}\right),$$
$$LA = 170 \text{ kg/ min}.$$

Now the flow parameter is quite different:

$$\frac{LA}{GA}\sqrt{\frac{\rho_G}{\rho_L}} = \frac{170 \text{ kg/min}}{(100 \text{ mol/min})\,[(0.0184 \text{ kg/mol})]}\sqrt{\frac{18.4 \text{ g}/22.4 \times 10^3 \text{ cm}^3}{1 \text{ g/cm}^3}},$$
$$= 2.6.$$

This value is higher than the normal design range, and it may require packed towers of larger cross sections than used for the amines.

Thus the original specification, that the absorption of nitrogen closely imitate that of carbon dioxide, will be difficult to achieve. Because complexing the nitrogen will require a complex of high molecular weight, the nitrogen concentration in the absorbing liquid will be less than that for the carbon dioxide. This means that relatively more liquid will be needed to remove the nitrogen, which in turn suggests that the packed tower geometry will change. This is by no means a fatal flaw for nitrogen absorption. It does mean that the specifications for the two absorptions will be different.

2.5   Conclusions and the First Gate

Product development begins with the identification of customer needs. Although the customers may be individual consumers or large corporations, their needs are effectively explored through interviews with fifteen to twenty individuals. Once these needs have been collected and rank ordered, they are best evaluated by the product development team, which should include representatives of marketing, research, engineering, and manufacturing. This team must reach a consensus on how the needs are translated into preliminary product specifications. Developing these specifications can be guided by first writing out all chemical reactions, by estimating mass and energy balances, and by assessing the rates of key processes. This is the point where the scientists and engineers on the development team first have a major input.

The preliminary specifications will require revision. This revision is facilitated by using a standard "benchmark," which is frequently a competing product. The revised specifications must get a critical analysis to see whether they make sense. The team must reflect on whether the specifications excessively restrict the possible solutions, as they did in the road deicing example. In addition, the specifications should be consistent with corporate strategy, if possible building on existing corporate strength. Once the specifications are complete, we are ready to start seeking ideas for their achievement.

The first management review of the product's development takes place at this point. Such a review is sometimes called a "gate," to suggest that the review will decide whether or not to continue the project. The core team will prepare both a written report and an oral presentation. Normally, the audience will be senior level managers, often well above the immediate supervisors of the core team.

The decision on whether to continue development will, at this early stage, usually be positive. After all, the core team will probably have been together for only a few weeks. Moreover, the impetus for the project may have come from very senior managers, who may be unwilling to see their ideas abandoned without careful scrutiny.

In fact, management studies suggest that this early review tends to be too casual, and that its conclusion should more frequently be negative. These management studies point to the high cost of canceling projects late in the product design procedure, and to the importance of quick checks on a spectrum of alternatives. The studies urge being carefully critical at this early stage.

We believe that there are two ways in which chemists and engineers can aid this early decision. The first way is to be ruthlessly objective, but without being destructive. This is a skill that comes naturally to some; others become easily involved in championing a particular perspective.

The second way in which chemists and engineers can be effective is in teaching management about the science behind the specifications. For example, we must explain that freezing point depression depends on the mole fraction of ions, and not on the mass of chemical. That is why we have so stressed specifications in this chapter. It is through specifications that those with scientific training can most help to decide whether product development should be continued or stopped. Product specifications are the key to this first management gate.

FURTHER READING

Astarita, G., Savage, D. W., and Bisio, A. (1983). *Gas Treating with Chemical Solvents*. Wiley, New York, ISBN 0471057681.

Atkins, P. W. (1998). *Physical Chemistry*, 6th edi. Oxford University Press, Oxford, ISBN 0198501021.

Bard, A. J. and Faulkner, L. R. (1980). *Electrochemical Methods: Fundamentals and Applications*. Wiley, New York, ISBN 0471055425.

Beckett, S. T. (1994). *Industrial Chocolate Manufacture and Use*, 2nd ed. Blackie, London, ISBN 0751400122.

Fogler, H. S. and LeBlanc, S. E. (1995). *Strategies for Creative Problem Solving*. Prentice-Hall, Upper Saddle River, NJ, ISBN 0131793187.

McGrath, M. E. (1996). *Setting the Pace in Product Development, A Guide to Product and Cycle-Time Excellence*, rev. ed. Butterworth-Henemann, Boston, ISBN 075069789X.

Rosenau, M. D. Jr., Griffin, A., Catellion, G. A., and Anschuetz, N. F. (1996). *The PDMA Handbook of New Product Development*. Wiley, New York, ISBN 0471141895.

Sato, K., Arishima, T., Wang, Z. H., Ojima, K., Sagi, N., and Mori, H. (1989). Polymorphism of POP and POS. *JAOCS* **66**, 664–674.

3

Ideas

Once we know the specifications for our target product, we need some good ideas. In fact, all we really need is a good one. Finding this idea is sometimes discussed by reference to the children's fairy tale, "The Frog Prince." You will remember the story: a somewhat vain princess who is walking in the woods promises her hand in marriage to a frog in return for some simple service. The frog performs the service and then shows up to claim his bride. The distraught princess submits ungraciously and then is astonished to discover that, after she kisses him, the frog turns into the prince of her dreams. Freudian psychologists have enjoyed interpreting this story.

For us, the story is the antithesis of product design. The chance of our beginning with one frog of an idea and making it into a prince is remote. Our chance of success is much greater if we can screen a large number of ideas. In terms of the fairy tale, we need to behave like a more modern princess, and kiss a lot of frogs. Exactly how many if we want to find a prince? Although estimates in different businesses vary, most experienced product developers suggest we need around 100 ideas to find one winner. We need to kiss 100 frogs to find one prince.

This large number of ideas can come from a variety of sources. Many ideas come from individual persons. These persons include customers, competitors, consultants, and members of the product development team. As explained in Section 3.1, generating these ideas should be as free and unrestricted as possible. This is the time for craziness, for off-the-wall notions, and for asking "What if?"

In some cases, the ideas generated by all these groups will be insufficient. In these cases, we will often seek additional chemical ideas sparked by the methods described in Section 3.2. We can look at natural products, checking if folk medicines are really effective. We can use the automated synthesis and analysis of combinatorial chemistry to explore many more compounds than has been possible with traditional synthetic techniques. These methods will increase the chemical possibilities.

Once we have generated a large number of ideas, we will want to winnow the best ideas from the large number assembled. Typically, we want to start with perhaps 100 ideas and wind up with five. We suggest doing this in two stages. As described in Section 3.3, we will first sort the ideas, removing redundancy. We will drop those groups of ideas that are inconsistent with our corporate strategy, or that do not build on our corporate strengths. We will nervously drop ideas that seem pure folly, knowing that some of these may contain seeds of innovation. This sorting of ideas will normally leave perhaps twenty survivors.

The second method of pruning our ideas, described in Section 3.4, is a more aggressive screening. This screening tries to cut the ideas from twenty to around five. We feel that one good route for this step is to use a concept screening matrix. In this procedure, we evaluate the general characteristics of each of our ideas – chemical understanding, engineering requirements, and so on – and compare weighted averages of these evaluations. The result of these screenings will be strong, practical ideas, any one of which should work.

3.1   Human Sources of Ideas

The consensus in product design is that we will normally need between twenty and 200 ideas to get one winning product. These estimates vary with the particular industry. Du Pont suggests that they need around 300 initial concepts to get one commercially viable product. 3M feels they can develop a winning product from only ten ideas. Zeneca and Pfizer suggest that they need around 100 ideas per success. Although these different estimates certainly reflect differences in what is meant by a new idea, the conclusion is clear: we will need a lot of new ideas.

Generating many ideas requires answering two questions. First, who are the sources of our ideas? Second, how do we get these sources to give us ideas? Exploring these questions is the subject of this section.

SOURCES OF IDEAS

Because we need so many ideas, we will accept them from any sources we can find. However, because some groups will be willing to spend much more time trying to help us, they will usually be our main sources.

One major source is the product development team itself. This team normally will have representatives who have made, used, and been frustrated by the existing products which we are trying to replace. They will be quick to see advantages and demerits of any new concepts. Their professional careers depend on success, so they have a large stake. They will be an excellent source.

The second group are the product's potential customers. These customers are those who will directly benefit from the new product's characteristics. The most important customers are the so-called "lead users." These customers frequently will already have tried to modify the product, improving its utility for their particular goals. Although such improvements may be of narrow scope, they often will identify new ways to meet specific needs. A related group to the customers are competitors, because both are interested in products that meet the same need. The competitors' marketing efforts may supply clues to their own plans.

The third area to look for ideas is the literature. The trade literature, and indeed trade shows, are the best source of information about current products beyond the product team and current customers. The archival literature, that is, the peer-reviewed publications of scientists and engineers, can provide the secrets of new products, but only if the product team has experts who critically read this literature. The patent literature is also valuable. Although many in management argue that patents are paramount, we feel that chemical patents are of limited value because they are often incomplete. We remember the anguished reports of chemists who, during the First World War, were trying to manufacture ammonia by using the Haber process. They concluded that however Haber made ammonia, it was not with the catalyst described in the patent.

Other persons who can be good sources of ideas are product experts, private inventors, and consultants. Experts are those with particular knowledge of the products that you want to make. Those retired from your organization or from a competitor may be especially helpful. Private inventors can be an important resource, especially for innovations that go beyond the boundaries of your current thinking. Dealing with private inventors is tricky: in many cases, their pet ideas may be impractical, but may spark different and important product ideas from the product development team. As a result, many companies keep careful written records of all dealings with private inventors to ensure all are clear on who invented what.

Consultants are the most difficult group to characterize because they are so diverse. Those who supply special services, like schemes for product development or innovation encouragement, seek to catalyze ideas from the organization's employees rather than produce ideas of their own. They can be valuable in limited doses. University professors can be frustrating consultants. The product development teams usually forget that professors are scholars, interested in truth and in education. They are the intellectual descendents of the monks who guarded society's knowledge. As such, they are often excellent critics but poor innovators.

COLLECTING THE IDEAS

The most direct and most effective method of collecting ideas is to ask the various groups listed above to write ideas down and send them in. Writing them down is important, for writing forces an objectivity that can spark improvements. If the ideas that we seek depend on chemical processes, a flow sheet can help. If the ideas include chemical synthesis, then guesses about synthetic routes and mechanisms are helpful. The ideas collected in this way form the core of our product ideas.

Asking the team, the customers, and a few experts for written submissions rarely will generate the hundred or so ideas that we will normally need. To get more, we will commonly assemble three or more groups of five to eight persons, and ask them to suggest still more ideas. Each group should have a formal leader who runs the session. Such a "brainstorming" session will work best under a few rules:

1. *Use a common format.* Have all groups cover the same topics in more or less the same way.
2. *Generate ideas freely.* Do not be worried that some ideas have problems.

3. *Eschew ownership.* Ignore which idea is whose, or whether the suggestion is competent.
4. *Encourage eccentricity.* Do not squash weirdness, even if it suggests the impossible.

Although it may take thirty minutes to get started, most groups will be able to generate ideas under these rules for an hour or two.

During idea generation, the group should keep a written record of its progress. Although the leader can sometimes do this, a separate scribe is usually better. The scribe often keeps his or her record on large sheets of paper which are then posted around the room. These posters are an enormous stimulant, because the group can refer back to old ideas when anyone thinks of improvements.

However, all brainstorming sessions will tend to stall, usually after about an hour. Although the group's productivity will drop, its creativity may actually rise, because the obvious avenues have been exhausted. Thus, most sessions probably should be kept going somewhat past the point where the group's members want to stop. To keep them going, we find four stimuli are especially useful:

1. Invite criticism of ideas generated by other routes. This produces comments such as "That's stupid, but if we went to an aliphatic side chain, then. . . ."
2. List all assumptions made in the specifications; then dismiss each in turn. This sparks statements such as "Well, I guess we don't really need a solvent if we use an aerosol. . . ."
3. Use analogies. Everyone remembers that heat conduction and mass diffusion are described by the same mathematics, but fewer remember that free energies can sometimes be changed by pH instead of temperature.
4. Probe opposites. This leads to comments such as "Reverse osmosis uses a water-selective membrane to separate a lot of water from a little salt in sea water. What if we use a salt-selective membrane?"

The purpose of these stimuli is to restart the group, encouraging their creativity in less explored directions.

PROBLEM SOLVING STYLES

In this brainstorming, we will probably see the emergence of different problem solving styles. Extensive psychological research, especially that by Kirton, suggests that it is useful to divide intellectual development into two styles, adaption and innovation. Adaption is problem solving that uses existing or closely related technology. Innovation is problem solving that uses apparently unrelated information. The choice in the psychological literature of the word "innovation" is unfortunate, for this word in the vernacular is often taken as a synonym for "creativity." Innovation may be creative, but adaption can be equally creative.

Problem solving styles are often evaluated by psychological testing. More simply, you can guess your own problem solving style by considering how your contribution to a project might be seen. If you are called a "nerd," a "drone," or a "bean counter," then you are probably an adaptor. If you are called a "flake," a "weirdo," or a "loose cannon," you are probably an innovator. We need both adaptation and innovation. How much of each depends on our objectives, for example, on the business we are in.

Not surprisingly, successful professionals within a single discipline tend to have similar styles, but different professions encourage different styles. For example, at one extreme, accountants are adaptors: the last thing you would want is an innovative tax accountant. At the other extreme, successful entrepreneurs are often innovative. Other professions fall in between these limits, in more or less the order expected: chemists are more innovative then engineers, who in turn are more innovative than elementary school teachers. Please remember than innovative is *not* a synonym for creative.

We also believe that there is an additional style beyond those of adaptors and innovators – that of a product "champion." A champion wants the product to work, whether it seems to make sense or not. He is often less deeply trained and less objective than either adaptors or innovators. He just wants the product to be successful. While he knows that a successful product will make money and advance his career, the champion also believes in the product, with an almost religious fervor.

We can make these three problem solving styles more understandable by giving three examples. As an adaptor, we choose Charles de Poitiers, the twelfth-century architect of the Cathedral of Notre Dame, in Paris. Charles worked for over 50 years, carefully extending the empirical knowledge of how to use stone to build higher and higher, "to the glory of God." Although the builders of this period knew no civil engineering, they came closer and closer to the limits of the stone they were using. Occasionally, these adaptors went too far: in 1248 AD, trying at Beauvais to increase the height of the nave 50% beyond Notre Dame, they built a cathedral that collapsed. Most of the time, however, they adapted and extended earlier successes to complete the magnificent structures that remain today.

Our example of an innovator is Michael Servitus, a fifteenth-century physician and monk with strong but unusual religious beliefs. Servitus became convinced that Christianity was a wonderful religion, except that there was no Trinity and Christ wasn't divine; otherwise the religion as practised in his day was fine. Like most innovators, he was sure he was right. He went to Geneva to convince John Calvin, who had him burned at the stake, a not unusual fate for innovators. They can have great ideas, but they can be a pain.

A good example of a champion is Lorenzo de Medici, "Lorenzo the Magnificent." Lorenzo died in 1492, making the date of his death easy to remember. More than any other single person, he was responsible for making Florence a powerful city-state, and hence sparking the development of nationalism. His motives and methods were certainly objectionable: he lied, cheated, and murdered, probably at least partly for personal gain. But he was important as a champion of Florence, and was certainly partly responsible for her ascension in politics and the cultural renaissance she harbored.

Just as individuals and professions tend to have a problem solving style, so do corporations. Although this may be just a consequence of the most common

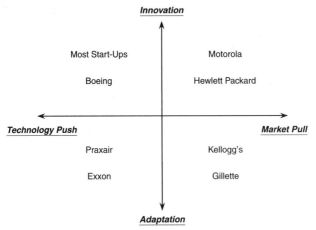

Figure 3.1–1. Corporate Problem Solving Styles. Different companies with different strategies inferred from market forces tend to develop different problem solving styles.

profession employed by the corporation, we suspect that it is often a deeper characteristic of the particular corporation. When we combine this style with the forces responsible for product development, we get company profiles like those suggested in Figure 3.1–1. Try plotting corporations of your own experience on these coordinates. We feel that most universities fall in the lower left-hand corner of this figure: they are pushed by technology but feature largely adaptive scholarship. We find a company like 3M especially illuminating. Their high development costs force them to be innovative, but they respond both to market pull and technology push. As a result, they would appear as a solid line across the top of the figure.

EXAMPLES OF UNSORTED IDEAS

The results at this point will be long lists of raw ideas. Three examples are given in Tables 3.1–1 to 3.1–3. These three examples will be discussed in more detail in the later sections of the book. Table 3.1–1 suggests ways to do laundry more efficiently, either by redesigning current washing machines or by developing new types of detergents. Although some of these ideas seem bizarre, we were amused to learn that one of our graduate students saves time and money by running his laundry through the clothes dryer three times more often than through the washing machine. These laundry speculations will be used in Section 3.3 to illustrate how ideas can be efficiently sorted. Such sorting can reduce the number of ideas from perhaps a total of 100 down to twenty.

The other two lists of ideas will be used in Sections 3.3 and 4.1 to illustrate how ideas can be more critically screened. Table 3.1–2 lists ideas for a new lithographic ink to replace an ink that contains a carcinogenic solvent and that requires large quantities of the same solvent to clean the presses. The new ink must offer an opportunity for reduced pollution. Table 3.1–3 is a partial list of ideas for treating

---

TABLE 3.1–1   Ideas for New Laundry Detergents Causing Less Pollution

1. Wash without soap.	26. Ultrafilter dirty water.
2. Throw the clothes away.	27. Imitate dry cleaning.
3. Use less soap and less water.	28. Get a new dry cleaning solvent.
4. Use a more effective soap.	29. Make a home dry cleaner that is sealed.
5. Add enzymes to detergents.	30. Use supercritical CO_2 for dry cleaning.
6. Add dead cells to detergents.	31. Use another supercritical solvent.
7. Add live cells to detergents.	32. Wash with Fuller's Earth.
8. Mop up dirt with particles.	33. Dry clean with chlorine-free solvents.
9. Use specific chemical interactions.	34. Grind the clothes up and remake them.
10. Improve the washing machine.	35. Recycle the surfactant by using a pH change.
11. Recycle the soap.	
12. Filter bigger detergent particles.	36. Recycle the surfactant by exploiting its cloud point.
13. Make larger micelles.	
14. Make emulsions out of the soap.	37. Make a detergent that precipitates on command.
15. Grow microbes on dirty clothes.	
16. Attach soap to particles, facilitating recycle.	38. A detergent that forms many phases.
	39. Wash clothes with dry shampoo.
17. Imitate dry cleaner agents.	40. Clean ultrasonically.
18. Use a fine adsorbent.	41. Shine with a UV light (to sterilize?).
19. Cook clothes under N_2.	42. Use pressure waves.
20. Air out clothes as washing substitute.	43. Cook clothes in high-pressure water.
21. Prevent soiling with antistatic coatings.	44. Freeze clothes; shake off dirt.
22. Wash until semiclean.	45. Calcine dry shampoo to make it pure.
23. Remove odor without removing dirt.	46. Use Fuller's Earth.
24. Wash with base, converting sweat compounds into soap.	47. Use ultrafiltration.
	48. Dry cleaning recycle is distillation.
25. Split objectives of clean, color-fast, and odor.	49. Flocculant aid for detergent.
	50. Adsorb detergent in clay.

Note: These speculations will be used in Section 3.3 to illustrate the sorting of ideas.

high-level radioactive waste containing ^{137}Cs. This isotope, produced in the manufacture of atomic weapons, is especially dangerous because it is water soluble. We will give details of the processes for dealing with the information in these three tables later; for now, we want only to illustrate typical results of the idea generation process.

3.2   Chemical Sources of Ideas

In many cases, the ideas resulting from the strategies given above for generating product ideas will show real promise in meeting the customer needs. The ideas may be devices for chemical change, such as better catalytic converters or cheaper kidney dialysis machines. They may involve particular chemical compounds, such as the Schiff bases that can selectively absorb nitrogen from natural gas. In some cases, however, the general strategies will not help much, because we do not know

TABLE 3.1–2   Ideas for a New Lithographic Ink with Reduced Solvent Emissions

1. Don't use a solvent.
2. Switch solvents.
3. Clean the press with robots.
4. Change the press.
5. Use an electrostatic ink.
6. Use a laser printer instead of the current design.
7. Change ink chemistry.
8. Recycle all of the solvent.
9. Clean the press with a high-pressure spray.
10. Extract the solvent from the rags used to clean the press.
11. Do the whole process in a clean room.
12. Isolate all equipment.
13. Clean the press less often.
14. Clean the press in a fume hood.
15. Print more checks at a time.
16. Mix the current solvent methylene chloride with other solvents.
17. Have each worker wear a self-contained breathing apparatus.
18. Use a solvent mixture.
19. Use a solvent that dissolves the ink.
20. The solvent in the ink should differ from the cleaning solvent.
21. Use a nonvolatile solvent.
22. Use partial cleaning of specific components of the press.
23. Steam clean the press.
24. Clean the press with air.
25. Put the press in a car wash.
26. Clean the press by brushing.
27. Clean the press by burning.
28. Make the lithography more like a jet printer.
29. Don't use checks.
30. Use a disposable press.
31. Use oil to trap the solvent.
32. Make checks by photocopying.

Note: More chemical details of the current ink are given in Example 3.3–2.

which chemical compounds we want to make. This may be especially true for pharmaceuticals.

When we do not know our target compounds, we can use chemical methods as a stimulus toward generating ideas. Three such methods are discussed here: natural product screening, random molecular assembly, and combinatorial chemistry. Natural product screening takes advantage of the rich variety of active chemical species present in nature. Plants, animals, fungi, lichens, marine organisms, and microbes are all rich sources of new chemical species with specific functions that could fulfill our needs. In random molecular assembly, molecular fragments are reacted in plasma to see if the resulting tar contains species which are, for example, pharmacologically active. Combinatorial chemistry uses robotics to provide a first

TABLE 3.1–3   Ideas for Treating High Level, Water Soluble Radioactive Waste

1. Store waste in bedrock.
2. Separate cesium with irreversible sodium titanate ion exchange.
3. Cesium extraction out of caustic solution.
4. Better process control for cesium tetraphenylborate precipitation.
5. Zeolite ion exchange.
6. Stabilize cesium tetraphenylborate precipitate with palladium catalyst poison.
7. Electrochemical separation of sodium.
8. Fractional crystallization to remove solvent.
9. Make a ceramic of waste.
10. Precipitate as cesium tetraphenylborate and vitrify quickly.
11. Electrochemical membrane support.
12. Potassium precipitation before cesium precipitation.
13. Cesium ion exchange with acid regeneration.
14. Build more storage tanks to hold waste.
15. Stabilize cesium tetraphenylborate precipitation.
16. Hollow-fiber extraction of cesium.
17. Dewater salt tanks for more storage capacity.
18. Precipitation at reduced temperature and storage as cesium tetraphenylborate.
19. Ion exchange on glass; vitrification.
20. Concrete formation of total waste (grout).
21. Simulated moving bed adsorption to separate cesium.
22. Inject in ground; vitrify *in situ* with nuclear explosives.
23. Concentrate cesium in microorganisms.
24. Adsorption in sodium titanate.
25. Regenerable ion exchange.
26. Cesium extraction in acidic solution.
27. Magnetic particles that adsorb cesium.
28. Electrochemical nitrate destruction.
29. Fluidized bed ion exchange.
30. Cesium separation and concrete formation (grout).
31. Cesium absorption in $MnFe(CN)_6$.
32. Flocculate tetraphenylborate precipitate.
33. Salt washing plus fractional crystallization.
34. Electrically regenerated ion exchange.
35. Reversible adsorption by using crown ethers on inert substrate.
36. Reduce explosion risk for cesium tetraphenylborate precipitate.
37. Use smaller size of cesium ion for separation.
38. Total vitrification of waste.
39. Make storage tanks safer.
40. Separate interstitial liquid, evaporate to dryness, and vitrify.
41. Buy additional benzene release permits.
42. Precipitate potassium before cesium.
43. Selective crystallization.
44. Properly designed ion exchange processes.
45. Properly designed precipitation processes.
46. Alternate precipitation chemistry.
47. Countercurrent exchange with sodium titanate.
48. Electrodialysis.
49. Salt dehydration and vitrification.

Note: This partial list of submitted ideas is used in Section 3.4 to demonstrate the concept-screening matrix.

pass at screening thousands, even millions, of compounds which may have the desired product properties. The chemical compounds discovered by these three methods can provide ideas for those working on the project. A synopsis of these methods is given in the following paragraphs.

NATURAL PRODUCT SCREENING

The first route to new chemical ideas is to look for possible sources in nature. During the past century, this has been a major source of complex chemical species with higher molecular weights. The pharmaceutical industry has benefited especially from mimicking nature: aspirin and opium, quinine and colchinine, caffeine and codeine all first came from natural sources. Other industries that rely on specific chemical activity have also benefited from natural products or their analogs. For example, stevioside is a low calorie sweetener derived from *Stevia rebandiana*, commonly called cua-hê-hê (sweet herb). Sunillin is a plant-based antifungal pesticide. A final, curious, example of a natural product analogy is the material for bulletproof jackets. This is based on a synthetic polypeptide, an alternating chain of glycine and alanine – particularly strong because of the small side groups and the well-ordered packing that results. This is exactly the structure of the fibers of spider's silk, one of the strongest fibers in nature.

There are three ways in which natural products may be used to produce active chemical species:

1. If the active ingredient is expensive or impossible to synthesize, it may be isolated directly from an organism. Vincristine, a highly effective treatment for childhood leukemia that is isolated from the Madagascan periwinkle, is one example.
2. A precursor may be isolated from a natural product and then used as a building block for a more complex molecule. Diosgenin, the active molecule in the first oral contraceptive, was originally synthesized entirely in the lab, but its economic production depends on a suitable precursor extracted from the Mexican yam (*Dioscorea floribunda*). A yam would seem an unlikely candidate to be responsible for a social revolution on the scale of birth control!
3. The active ingredient may be identified in a natural product, but then used as a model for a chemical synthesis of an identical or similar molecule. Reserpine, a drug used to treat hypertension, was first identified in the Indian snakeroot (*Ravolfia serpentina*), used in traditional Ayuvedir medicine in India as a tranquilizer, but is now produced entirely synthetically.

Although the first group is historically the most significant, the third category is the target of most modern investigations, particularly by use of microorganisms.

The exploitation of microorganisms merits further description. We start by seeking samples of new microorganisms. Fewer than 0.1% of bacteria and fungi present in the soil have been tested; yet penicillin, cyclosporin (the immunosuppressant that makes transplant surgery possible), and lovestatin (the world's leading drug for lowering cholesterol levels) are just a few of the products that were derived from fungi. One way to find these is to send a graduate student armed with a trowel off to a tropical country for a few weeks. The student will return with bags of soil samples that can be cultured to seek new, unusual microbial species. Alternatively, we can look close to home: the mold from which current penicillin cultures are descended was found growing on a cantaloupe in the garbage of a supermarket in Peoria, Illinois.

As another approach, we can begin with a culture that we already know produces active chemical species. We then stress the microbe either by chemical treatments or, more classically, by high doses of radiation. These doses should be so high that they will kill most of our culture. We will then grow the few survivors to see if the stress of radiation has rearranged their DNA beneficially.

We will often test these new microorganisms by culturing them within an existing colony of the target pathogens. In a few cases, we will find that our new microbes will kill target pathogens growing close by, presumably because the new species is producing a toxic chemical. Seeing such behavior is said to have been the stimulus which led Alexander Fleming to the discovery of penicillin. Thus we can discover which of our candidate microbes are producing toxins of value.

We next need to determine the chemical structures of these toxins. To do so, we will normally depend on a combination of chromatography, mass spectrometry, and nuclear magnetic resonance. Hopefully, we will discover the chemical structure in sufficient detail so that we will be able to compare synthesizing the target species by chemical methods or by fermentation with our new microorganism. If we decide on fermentation, we can try to dramatically increase the production (the titer) by inducing mutation in our new mircoorganism.

In addition to microorganisms, we can seek active chemical species in plants. This rich source of ideas is hardly touched. Although there are on the order of half a million species of flowering plant, only 5000 of these have been extensively investigated for active chemicals, and most of those originate from temperate regions. The potential of aquatic organisms has also recently started to be realized: discordomolide (a powerful immunosuppressant) and marinovar (an antiherpes agent) have recently been extracted from a Bahamanian sponge and a Californian marine bacterium, respectively.

The testing of microorganisms and plants has been dramatically accelerated by advances in pharmacology, which allow assays for compounds that can selectively bind to receptors associated with known physiological events. For example, in a rapid automated test, the National Cancer Institute now routinely screens extracts from natural sources against an array of up to sixty distinct human tumor and cell lines. Promising samples are then obtained in larger quantities. These specific bioassays have developed in parallel with high throughput screening based on microelectronics, robotics, and advanced spectroscopy. Such integrated systems can screen thousands of samples daily and effectively pinpoint those with pharmacological, agrochemical, or other industrial utility.

A final approach to natural product screening is to investigate the traditional uses of natural products, primarily as medicines. This is not a new method. In 1785, William Wittering discovered that ingestion of dried foxglove (a member of the genus *Digitalis*) eased dropsy, a condition caused by the heart's failure to pump properly. He reported, "I was told this had long been kept secret by an old woman in Shropshire, who had sometimes made cures after the more regular practitioners had failed." Today two components – digitoxin and digoxin – are prescribed to cardiac patients to regulate and strengthen their heartbeats.

At present, this method centers on conserving the knowledge of traditional healers, who may be remnants of dying cultures in Third World countries. For example, flavanone is a topical anti-inflammatory extracted from tree bark in traditional Samoan medicine. Another example, the adrenergic blocker yohimbine, is extracted from *Pavsinystalia johimbe*, the source of a traditional aphrodisiac.

Thus natural product screening can provide a rich source of chemical ideas.

RANDOM MOLECULAR ASSEMBLY

In this simplest of the three methods, we have a vague idea of the type of molecule that we want to make, but we are uncertain about its chemical structure. For example, we may want to make modified penicillin with some new substituents. In this case, we simply take an existing penicillin and some chemical species that contain the core of possible substituents. We do not even need to insist on the purity of the starting materials. However, if possible, we dissolve all these in a reasonably homogeneous solution.

We then inject this solution into a plasma. Such a flame is simply an ionized gas with a high concentration of free radicals. The radicals cause a plethora of chemical reactions, and, under some conditions, yield a dark, viscous tar. This tar, containing fragments of our original molecules jammed together in near-random configurations, is our product.

We must next test our tar for pharmacological (or other) activity. In many cases, we will have a particular disease in mind. For example, we may be interested in a gonorrhea bacterium that has developed a high resistance to penicillin. We begin by growing this bacterium in a Petri dish containing a layer of Agar gel, which is filled with nutrients for the bacterium.

We then put a drop of our plasma-produced tar in the middle of the bacterial colony. If the tar has no pharmacological activity, the colony will continue to grow unchecked. In some cases, the colony's growth will slow, but not much differently than if the colony were challenged by the original unreacted solution. But in a few cases, a drop of tar will kill the colony, destroying the microorganisms. Then we know that our tar contains some promising molecules.

We must next discover what these molecules are. To do so, we normally will make a solution of our tar, and separate this solution by high-pressure liquid chromatography. This chromatographic method uses a column packed with 5- to 10-μm spheres, and so will have a large pressure drop – over 10^4 kPa – at any reasonable flow rate. The packing in the column does not matter greatly at this stage: we ask only that the packing give a crude separation of the tar into perhaps ten fractions.

We then test each of these fractions for pharmacological activity. Although the activity may be reduced by dilution, we will normally find that it is isolated in one or two fractions. We will then further separate the most active fractions, using a different column which will get closer to separating the fraction into individual chemical species. When we feel we have these species isolated, we identify their chemical structures, usually with a combination of mass spectrometry and nuclear magnetic resonance.

The fascinating part of this method is our substitution of the brute force of the free radical-filled plasma for any chemical insight. We are not thinking about chemical mechanisms at all; we are simply trying to stick chemical fragments together. Once they are together, we are not thinking about how we would make them; we are just trying to identify which ones will kill microbes. Only when we identify the killers do we determine the chemical structures involved.

These plasma-generated chemical species are essentially chemical ideas found randomly. We can also generate other promising chemical species robotically, using combinatorial chemistry. This is described next.

COMBINATORIAL CHEMISTRY

This is the chemical equivalent of using a sledgehammer to crack a nut. Though the sledgehammer may be inelegant, it is unquestionably powerful! Very often we are faced with a chemical problem we know we want to solve: a drug to attack a specific protein, a catalyst to speed a known reaction, or a poison specific to a microbe. We might, however, have no idea how to solve this problem. There are thousands of potential solutions out there; we have no chance of testing them all other than via combinatorial chemistry.

The core idea behind combinatorial chemistry is to identify possible active ingredients or molecular fragments and to test all of them and in all possible combinations. Clearly, both the combining of ingredients and testing of activity need to be automated for this to be practical. The method has lent itself particularly well to biochemical problems. Most biological molecules are by their nature sequences of a limited range of alternatives. DNA is a repeating sequence of the four bases, A, C, G, and T. Proteins are a linear sequence of twenty naturally occurring amino acids. Naturally occurring fats are almost all triglycerides: only the chain length and degree of saturation of the fatty acids vary. Clearly it is a conceptually easy matter (though potentially very painful) to synthesize all possible DNAs, polypeptides, or fats and to test them on the biological systems of interest.

As an example, imagine that we want to investigate the efficiency of hexapeptides for affinity for the μ-opioid receptor. Houghton and co-workers identified a potential library of 52,128,400 hexpeptides which might have the desired affinity. Even for a robot this represents a lot of work. They therefore decided to structure the problem. First they tested 400 alternatives, in which only the first two amino acids were varied. The most efficient of these were then taken as fixed, and successive amino acids are used to optimize efficiency; that is, a further four tests of twenty alternatives each. This strategy thus paralleled the method of steepest descent used in chemical process optimization. Encouragingly, the molecule identified by this particular search has the sequence occurring naturally in proteins that stimulate this receptor. This demonstrates the power of the combinatorial method.

TABLE 3.2–1   Three Routes to Chemical Ideas

Parameter	Natural Product Screening	Random Assembly	Combinational Chemistry	Remarks
Typical Starting Information	Potions from folk medicine	Similar chemicals of known structure	Similar chemicals of known structure	Note that we must have some chemical knowledge to begin
Chemical Synthesis	None	Random assembly of known fragments	Planned assembly of known fragments	We may not know what we have made
Trials for Efficacy	Use entire potion	Use entire product mixture	Use each known product	We will discard most of the chemical species present
Chemical Analysis	Identify active ingredients	Identify active ingredients	None	We know what we have made only in the third case

Note: Natural product screening is well developed and combinatorial chemistry is evolving rapidly. Random assembly is less often used.

At the same time, this example shows that even within the combinatorial method, there is considerable skill and invention involved in identifying the key parameters to vary systematically. That is true for this simple hexapeptide, and even more so for more complex cases. In these more complex cases, we often discover synergistic effects between molecules. We find important variations caused by concentration, temperature, pressure, and pH. Judgment and chemical understanding remain critical in effective use of the combinatorial method.

The combinatorial method remains in its infancy. It is effective for catalyst screening. It has already gained great exposure for identifying active polypeptides and DNA sequences. It has found a significant role in testing positive and negative synergies in drug cocktails. The true potential of the method is revealed when we remember that up to 1991, about eleven million compounds had been identified in the chemical literature. By combinatorial methods, that number can be made by a simple machine (a so-called librarian) in less than a week.

Before reviewing an example, we pause to compare these three methods. A synopsis of their chief characteristics is given in Table 3.2–1 for the case of pharmaceuticals. In every case, we begin our search for ideas with some chemical information. For example, we know of a potion from folk medicine, or of drugs of limited efficacy. Often, we use this information as a starting point for synthesis. We then test the results of our synthesis to see if we have uncovered pharmacological activity. When we have this activity, we may need to analyze our reaction products to identify the source of the activity.

We illustrate these ideas with an application to fuel cell catalysis.

EXAMPLE 3.2–I   FUEL CELL CATALYSIS

An elegant example of combinatorial methods used in catalytic chemistry is given by Milhawk and co-workers. The problem is to optimize the composition of a catalyst for methanol fuel cells. High surface area Pt-Ru catalysts are the current technology, but waste about 25% of the fuel's energy. By chemical analogy, we want to consider the other platinum group metals Os, Ir, and Rh, as additives. The problem is to test efficiently many catalysts of different composition and with different combinations of elements.

How can this be done?

SOLUTION

To make these tests, Milhawk and co-workers built a modified ink jet printer to spray dots of mixed metal salts. They produced a 645-member electrode array, including the five pure elements, eighty combinations of two elements, 280 ternaries and 280 quaternaries. The dots were dried and reduced to the metals. To test each combination, Milhawk and co-workers used a fluorescent molecule that luminesces in acid but not base. (H^+ is produced as part of the fuel cell's catalytic cycle.) On testing, the most effective catalyst simply lit up the brightest.

The results are fascinating. A quaternary alloy, Pt(44)/Rh(41)/Os,(10)/Ir(5), was found to be the most efficient catalyst. It was significantly more efficient than the commercial binary, although the surface area was only about half. Other alternative catalysts were identified in different regions of both ternary and quaternary space. The most efficient ternary, Pt(62)/Rh(25)/Os(13), lies in a ternary region bounded by inefficient binaries, Pt-Os and Pt-Rh. We would not intuitively expect high activity from this ternary composition. These results could never have been achieved by conventional catalytic testing: the amount of work required would be just too great and the results show that a rational or intuitive approach would have failed.

3.3   Sorting the Ideas

Generating possible solutions to a need is fun, but it produces a hodgepodge of ideas. These ideas are incomplete. They are not so much like frogs which must be screened to find a prince; instead, they are like fragments of frogs. Somehow, we must compare these fragments and choose the most promising candidates.

The situation seems to us much like the screening of candidates for an open professorship. If we advertise such a position, we will get at least 100 applications. We want to narrow this number to about five, whom we wish to invite for an interview. We hope that at least one of the five will be acceptable. However, we do not have the time to carefully evaluate all the applicants. How do we go from 100 applications to five?

One possible strategy uses two separate stages. First, we could eliminate those who are unqualified. This could include candidates without a doctorate or without publications and research funding over a long career. We could eliminate recent PhD graduates from universities without much research effort. We could eliminate noncitizens or older applicants, though such actions could be illegal. We could eliminate those in a particular subspeciality, like polymeric materials, feeling that our department is already well represented in that area. We would use this strategy almost without thinking, just to cut the number of applicants down to perhaps twenty. These twenty we would carefully study to choose the five for interviews.

The situation here is similar: we will have perhaps 100 ideas, and we want to choose the best five. Just as we might do for faculty candidates, we will make this choice in two stages. First, we will sort and prune these ideas, which will normally reduce the number to around twenty. Second, we will use matrix-screening methods to choose the best five. The sorting is the subject of this section, and matrix screening is covered in the next section.

GETTING STARTED

The first step is simply to prepare a list of all the ideas. The list will almost certainly contain considerable redundancy, which is easily removed. The redundancy most often will occur because some ideas are more general and some are more specific. For example, in an effort to recover orange juice flavor during juice evaporation to make a concentrate, one idea could be:

"Remove the flavor from the vapor."

A second idea could be:

"Use a selective membrane to concentrate flavor from the vapor."

The second idea is a specific example of the first.

 The overall list will also contain some ideas that seem folly and that should be pruned. This pruning is much trickier than removing redundancy. Some ideas will simply be irrelevant, perhaps recorded incorrectly or not thought through in the turbulent brainstorming session. They are easy to drop. Some ideas may prove just plain wrong, perpetual motion machines in disguise. However, some ideas may be incorrect but still contain innovations that we do not want to lose. As a result, we suggest a bin of random thoughts that we keep aside from the rest of the sorting, and periodically recheck. Removing redundancy and folly will typically cut the number of ideas by about a third, from around 100 to about seventy. We need to organize these ideas further in an effort to get down to the targeted twenty.

"THE MATERIAL WILL TELL YOU"

The next step is to organize the ideas into categories, but the hard part is to know what form the categories should take. In this effort, the best guide is to remember what good writing teachers say: "The material will tell you." By this, the teachers mean that the structure will never be the same for any two sets of ideas. The organization of ideas for recovering orange juice may be as different from that for ideas for a reusable detergent as from the structure of an essay on "Macbeth." The ideas themselves must be the basis of any organization.

In many cases, we will find the organization of the ideas is obvious, and jumps out to all members of the core team. Be careful that the organization does not simply reflect the training of the core team. In one case, a team of mostly engineers insisted on organizing the ideas around unit operations. In so doing, they overrode team members trained in chemistry who urged organizations based on chemical mechanisms. The latter scheme would, in this case, have provided more insight.

The normal rules of outlining apply to this effort. First, you want to use around five main headings. These five headings should be roughly equal in importance. If you have many more headings, examine them carefully to see if some cannot be combined. Although sometimes they cannot, make the effort.

Second, subheadings should be special cases of each main heading. There should rarely be more than four subheadings; if there are, consider combining these further. There should never be only a single subheading; if there is, then the subheading should probably replace the main heading.

In the outlining, we should remember a guideline sometimes taught in elementary school, called "The Rule of The Table." The guideline argues that an outline is like a group of dinner tables. Each heading is like the top of the table, covering everything under it. The subheadings are like the table's legs. Three or four legs are the best number for a table. A few good tables do have two legs (harvest tables), but very few good tables have six or eight legs. Thus, says this guideline, make the sections of your outline be like tables. Although this may sound simplistic, we have found this rule helpful.

Once the outline is made, it should be carefully edited. Most commonly, the editing will focus on three areas. First, the outline will often expose gaps. For example, in an evaluation of reducing pollution from airplane deicing, we might have a heading "Extract the deicer from water for recycle." This really should be a subheading under "Recycle the Deicer," which may not have been one of our original ideas. Such a heading could have subheadings like "Extract the deicer," which we got, and like "Adsorb the deicer," which we missed. We may decide to add this new "Adsorption" idea.

The second area for editing is to prune unpromising or politically impossible ideas. For example, one group of ideas may call for inventing a new type of chemistry, and we may have little in-house chemical skill. In this case, this group of ideas should probably be dropped. Similarly, our company may make centrifuges, and we may have ideas for membrane separations. Pursuing the membrane ideas could be a completely new direction for our company, and hence imply deciding to shift corporate strategy. Such a shift probably is beyond the mission of our particular product development team.

The third area for editing is to acknowledge different patterns of thinking in different individuals. Some persons cannot organize ideas in a linear structure

TABLE 3.3–1   Ideas for an Adhesive for Wet Metal

1. Wipe the metal surface with a cloth (F)
2. Change the metal's composition
3. Change to a new adhesive (V)
4. Make the adhesive water absorbing
5. Use a plant that sticks to a ship (V)
6. Use a natural rubber
7. Electrostatically charge the metal
8. Put a magnet in the current adhesive (R-7)
9. Use a super glue (i.e., a cyano acrylate) (V)
10. Use a different resin (V)
11. Make a resin with a hydrophilic part
12. Treat the surface with zeolite
13. Use a zinc coating primer
14. Spray on a silicone coating
15. Use neoprenephenolic as the adhesive (R-59)
16. Invent an adhesive that reacts with water
17. Use a silica gel for surface treatment (F)
18. Choose a van der Waals bonding material
19. Try ionic bonding (F)
20. Use a water scavenger in the adhesive base
21. Treat the surface with alkali
22. Use corn starch (F)
23. Use an adhesive with a functional group that reacts with water
24. Use an isocynate with a water reactive part (R-23)
25. Inject acidic salt in the metal (F)
26. Use more adhesive (F)
27. Use a concrete cement
28. Choose a water catalyzed polymer
29. Choose an adherent with a water reactive part (R-23)
30. Use a water scavenging adhesive (R-20)
31. Add a catalyst to speed up the reaction
32. Invent coupling chemistry (V)
33. Adapt dental adhesives (V)
34. Try a polymer with a protective layer and heat to use
35. Use heat catalyzed polymer (V)
36. Coat the surface before applying the adhesive (V)
37. Invent an adhesive that reacts with metal
38. Welding with a laser (F)
39. Replace the metal (F)
40. Solder (F)
41. Use a sugar solution (V)
42. Use a reversible glue (F)
43. Apply a vacuum adhesive (V)
44. Use an adhesive developed for the bathroom (V)
45. Use candle wax (F)
46. Try water based adhesives (R-11)
47. Use natural rubber (R-6)
48. Use spider web (F)
49. Use asphalt
50. Eliminate metal from cars (F)
51. Use rope to tie up the metals
52. Use bubble gum (F)
53. Make plastic or fiberglass cars (F)
54. Don't use cars (F)
55. Use flower tapes (F)
56. Blow dry the surface
57. Use toluene as the base solvent (R-4)
58. Get water resistance by using nitrocellulose/polyisobutylene
59. Use a phenolic group
60. Use a zipper (F)

like that used here, and may prefer circular outlines or pie charts or some other structure. Some may prefer decision trees. In particular cases, organizations may have adopted particular development schemes. Any of these can work. Which you use does not matter. Just use those that let you cut the number of ideas from your original 100 down to twenty.

EXAMPLE 3.3–1   ADHESIVES FOR WET METAL

A product development team wants to produce a new group of adhesives that will stick to wet metal surfaces. Such an adhesive would have value in the automotive industry. The team has produced the ideas shown in Table 3.3–1.

Sort these ideas into a coherent organization.

TABLE 3.3–2 Sorted Ideas for Wet Metal Adhesives

I. Improvements in Existing Adhesives
    A. Choose a van der Waals bonding material (18).
    B. Use an adhesive + a functional group that reacts with water (23).
    C. Add catalyst to speed up the reaction (31).
    D. Coupling chemistry (32).
    E. Apply a vacuum adhesive (43).

II. Water-Absorbing Adhesives
    A. Make the adhesive water absorbing (4).
    B. Make a resin with a hydrophilic part (11).
    C. Invent an adhesive that reacts with water (16).
    D. Use a water scavenger in the adhesive base (20).
    E. Choose a water catalyzed polymer (28).

III. Surface Treatments
    A. Treat surface with zeolite (12).
    B. Use zinc coating primer (13).
    C. Spray on a silicone coating (14).
    D. Treat the surface with alkali (21).
    E. Try a polymer with a protective layer and heat to use (34).
    F. Blow dry the surface (56).

IV. New Innovations
    A. Change the metal's composition (2).
    B. Use a natural rubber (6).
    C. Electrostatically charge the metal (7).
    D. Invent an adhesive that reacts with metal (37).
    E. Get water resistance with nitric cellulose/polyisobutylene (58).
    F. Use a phenolic group (59).

V. Curiosities
    A. Use a concrete cement (27).
    B. Use asphalt (49).
    C. Use rope to tie up the metals (51).

Note: The numbers in parentheses refer to Table 3.3–1. The ideas in boldface are felt to be the most promising.

SOLUTION

To begin to sort these ideas, the team considered which seemed folly, vague, or redundant. These are labeled as F, V, or R in Table 3.3–1. The remaining ideas were grouped under the five rough headings shown in Table 3.3–2. These include improving existing adhesives and treating the metal surfaces. Note the final heading, a collection of ideas that the team felt had merit, but that are hard to imagine in practice.

Interestingly, in this case, the team felt that it found no competitive advantage for a new product. This product development was abandoned.

EXAMPLE 3.3–2   REUSABLE LAUNDRY DETERGENTS

Our start-up company has raised considerable resources to seek pollution preventing, environmentally benign chemical technologies. Without many specifics,

TABLE 3.3–3   Sorting Ideas for Reuseable Laundry Detergents

  I. New Soap (4, 9, 11, 37)
    A. Chemistry
       1. Base (24)
       2. Powders (8, 17, 18, 32, 39, 46, 50)
       3. Biochemical (5, 6, 7, 15)
       4. Emulsions (14)
    B. Easier to recycle (48)
       1. Size (12, 13, 14, 16, 26, 47)
       2. Temperature (14, 36, 38, 45)
       3. pH (35)
       4. Other chemistry (49)
 II. New Washer Design (3, 10)
    A. More mechanical energy
       Ultrasonic and pressure waves (40, 42)
    B. Thermal energy
       1. Cooking (43)
       2. Freezing (44)
    C. Light (41)
III. Improved Dry Cleaning (1, 27, 28, 48)
    A. Altered equipment
       1. Sealed (29)
       2. Supercritical (30, 31)
    B. New solvents (33)
       1. Gases (19, 20)
       2. Liquids (30)
IV. New Directions
    A. Disposable clothing (2, 34)
    B. Soil resistant clothing (21)
    C. Altered mores (22, 23, 25)

Note: The numbers in parentheses refer to the ideas given in Table 3.1–1.

the company's prospectus promises "reusable detergents." These should be more attractive ecologically than many existing "environmentally friendly" detergents, some of which have high sodium hydroxide concentrations. A few consumer surveys generated product ideas like those shown in Table 3.1–1.

Sort these ideas so that we can begin to choose those that are most promising for development.

SOLUTION

The ideas in the earlier table break into the four groups shown in Table 3.3–3. The numbers shown in parentheses refer to the particular ideas in Table 3.1–1. These four groups represent different regions on the intellectual triangle in Figure 1.2–1. For example, those under heading "I.   New Soap" is close to the "chemistry" corner of the triangle. "I.A.   Chemistry" is nearest this corner. "1.B.   Easier to Recycle" also depends to some extent on chemical engineering. The heading "II.   New Washer Design" is going to be dominated by mechanics, and be relatively independent of chemistry.

This organization of ideas shows immediately that our new company needs to choose between several, very different strategies. The first heading implies making a new soap, and so makes sense for a company like Proctor & Gamble, who already make conventional detergents. It would be a major new initiative for a company like Whirlpool, whose current business centers on making washing machines. Whirlpool has already explored the topics under "II.   New Washer Design." Interestingly, the washing machines available in Europe are dramatically more expensive and slower than those available in North America. However, European machines use both less water and less detergent, and hence cause less environmental intrusion.

EXAMPLE 3.3–3   A POLLUTION PREVENTING INK

A printing company prints personal checks with a lithographic ink containing the carcinogenic solvent methylene chloride (CH_2Cl_2). Workers at this company also clean the presses by wetting a shop rag with the same solvent, and scrubbing down the press. This procedure works well. The trouble is that much of the methylene chloride evaporates and so risks workers' health and censure from the environmental authorities. Also, the soiled rags have recently been reclassified as a hazardous waste, so that the cost of their disposal almost equals the cost of buying the solvent in the first place.

The company clearly needs to use a different ink, one that has less negative environmental impact. Some ideas for this ink are shown in Table 3.1–2. Sort these ideas to identify those most worth pursuing.

SOLUTION

The ideas easily break into four groups, as shown in Table 3.3–4. Again, the numbers in parentheses refer to the original sequence of ideas, which in this case are in Table 3.1–2. The first group in Table 3.3–4 involves changes in the printing presses. Because the company does not want to make the enormous capital investment involved in changing the presses, this group is deferred until other alternatives have been explored.

The second group of ideas involves either containing the solvent or using a different solvent. These ideas are the easiest to implement, and hence the most tempting for further development. The third group of ideas implies the invention of a new ink, a more major effort than the substitution of a new solvent. We will explain this option more in Section 4.1.

The final idea, "Don't Use Checks," may initially seem foolish; but consider the explosion in electronic money transfers. The company may decide that electronic data processing, which replaces handwritten checks, is like the automobile which replaced the horse-drawn buggy. If so, then printing checks may be like making

---

TABLE 3.3–4   Sorting Ideas for a Pollution Preventing Ink

I. Improve Current Printing
   A. Change press (4)
      1. Isolate press (3, 11, 12, 14, 17)
      2. Use laser printer (6)
      3. Use photocopying (32)
   B. Change cleaning
      1. Less often (13, 15)
      2. Other solvents (9, 23, 24, 25, 27)
II. Use a New Solvent
   A. Change CH_2Cl_2 operation
      1. Recycle (8)
         a. Extract (10)
         b. Spin dry (new)
      2. Burn (27)
      3. Freeze (new)
   B. Replacement of CH_2Cl_2 (2, 20)
      1. Nonvolatile solvent (21)
      2. Oil as solvent (31)
      3. Solvent mixtures (16, 18)
III. Solvent-Free Ink Chemistry (1, 7)
   A. Electrostatic ink (5)
   B. "Solvent that dissolves ink" (19)
IV. Don't Use Checks (29)

Note: The numbers in parentheses refer to the ideas suggested in Table 3.1–2.

---

buggy whips. Thus this fourth idea should be carefully considered in the idea screening process that is to follow.

3.4   Screening the Ideas

Organizing the fragments of ideas that we have generated can greatly clarify our thinking, reducing the number of concepts for more quantitative consideration. For simple product designs, the idea sorting may suggest only one or two strong ideas for new products. If this is the case, we can skip directly to the "Selection" procedure described in Chapter 4.

In many cases, however, our search for new products will have generated a large number of fragments of ideas. For many industries, we have suggested that this will be around 100. Sorting through these ideas to remove redundancy and inconsistency with corporate goals will normally cut this number to twenty. We need to cut the number further, for we will not normally have the resources to make the quantitative calculations necessary for selection between all of these. We need a basis for qualitative judgments to further reduce the selection.

We explore one method for achieving this in this section. To compare different product ideas, we recognize that our comparison must include a wide variety of criteria. It will include purely objective questions, such as "Which of these two absorbents has a greater capacity?" and "Which battery has a greater power per mass?" Some objective comparisons will also involve cost. We will normally prefer a catalyst that has only half the activity but one tenth of the cost of our original benchmark.

At the same time, our comparison of different product ideas should often include more subjective criteria. We will be less interested in a new nonwoven fabric if it is less "wearable." We will normally prefer a marginally more expensive product that we judge is "safer." In more complex cases, we will be making compromises between two conflicting criteria. For example, we may be hard pressed to decide between different home air purifiers whose performance and cost go up together.

STRATEGIES FOR IDEA SCREENING

There are clearly many possible strategies by which we can screen our ideas for product design. Our easiest approach to screening the ideas is to look at the headings in the outline, and choose the best candidate under each heading. We then use these candidates in the selection procedures in the next chapter. This simple strategy can work well if the product designs are simple extensions of existing technology. It does have the significant risk that the two best ideas will be under the same subheading. This strategy will have major risk if there are many, very different product designs.

A more effective strategy is to choose the most important factors by which we want to evaluate the product. These factors often will include at least some of the following:

1. *Scientific maturity.* We will prefer designs based on scientific knowledge that we already have and understand.
2. *Engineering ease.* We will prefer designs that imply straightforward engineering like that already used in established manufacturing.
3. *Minimum risk.* We do not want to take unnecessary chances. At least, we want to know what our chances of success are.
4. *Low cost.* We may want a rough estimate of the relative cost of our ideas.
5. *Safety.* We want to identify which products are inherently safer or more dangerous than our benchmark.
6. *Low environmental impact.* We will tend to choose products that cause less pollution.

Other important factors may be more subjective. Examples are "the product should be quiet," or "the product should be comfortable."

We next need to choose five or fewer of these factors that are most important. This choice is best made by consensus, with the entire core team working together. Asking the individual members of the core team for their choices and averaging these choices usually does not work well. In seeking this consensus, the members of the team need to be careful not to compete, not to feel that their chosen factors will be "winners" or "losers." We are not sure how to ensure this. We do recognize that some individuals have personalities that catalyze rational judgment, just as other individuals can precipitate polarization and win-lose arguments.

Once the key factors are identified, we need to assign weighting factors to them. These weighting factors should be normalized:

$$\sum_{i=1}^{n} \omega_i = 1, \tag{3.4-1}$$

where ω_i is the weighting factor of attribute i, which is one of a total of n attributes. In our experience, the core team will reach a consensus on these weighting factors more quickly than on the choice of the factors themselves. We suspect that this may be because the discussion about which factors are important implicitly includes a discussion of the relative weights.

With these weighting factors in hand, we now evaluate the key ideas in our outline on the basis of some scale. The easiest scale for most core team members ranges from a low score of one to a high score of ten. First, we assign an average score of five to our benchmark product, and we grade each of our product ideas relative to this benchmark. When we are through grading, we have a group of scores s_{ij} for each attribute i and each idea j. We then calculate the total score for each idea:

$$\text{Score}(j) = \sum_{i=1}^{n} \omega_i s_{ij}. \tag{3.4-2}$$

The ideas with the highest scores are those we will use for the next stage of product design, selection.

IMPROVING THE IDEA SCREENING PROCESS

The process described above often strikes those trained in engineering or chemistry as too easy and too qualitative. Even when relative scores seem reasonable, they are obtained with so little effort that they are suspect. We believe that this intellectual masochism may result from the study of subjects such as thermodynamics. These strategies for idea screening are certainly simpler than standard state chemical potentials or fugacity coefficients. As a result, many chemists and chemical engineers will concede that, "Yes, well, the concept screening process is interesting, but we cannot really take the results seriously."

We disagree.

We believe that the result of the concept-screening matrix should be taken seriously, though the simple procedure outlined above can be improved. We suggest three specific improvements. First, we urge a careful choice of the benchmark. In many cases, this benchmark will be an existing product with the greatest market share. In some cases, it may be what we expect as a new product from competitors. In other cases, it may be what we hope we can make as the best of the existing type of product. As a check on this choice, we may sometimes try to choose a different benchmark after a first round of assessments, just to make sure our first benchmark is best.

The second way to improve this concept-screening matrix is to check the core team's scores against those of other interested experts. One obvious group are other individuals in marketing who are outside our core team. Another group are the lead users of current products. In seeking these outside opinions, we must remember that the uninformed tend to be conservative: they are those who will not buy "genetically engineered tomatoes" but will buy "new improved hybrid tomatoes." Although we should seek these outside opinions, we should not accept them as gospel.

The third way we should improve the screening process is to make a sensitivity analysis of the weighting factors. Although the details of this analysis are beyond the scope of this book, we will essentially change the weighting factors within sensible limits to see if this alters our rank ordering of the ideas. In most cases, we will find little change. In a few cases, where change is dramatic, we should re-examine our selection criteria, for we have probably not explicitly considered all the major issues.

Although we believe that it has value, the concept-screening matrix is not a panacea, able to screen all available ideas in all situations. Some problems may come from the tendency to compress the scale of judgments. After all, if you give everything a score of 5, you are effectively abstaining from judgment. If all vote in a similar way, then that particular selection criterion has no value and should be dropped.

A much more serious problem is the assumption implicit in Equation 3.4–2 that everything can be scored and weighted linearly. This is approximately true only when the products are similar, changed in only minor ways. The assumption of linearity is generally untrue in three obvious cases:

1. *The criterion is binary.* For example, the product could be judged noisy or quiet, with nothing in between.
2. *The product will not work.* For example, the top-scoring product may depend on making an invention which may not be possible.
3. *The product changes the market.* This implies an innovation that is so good that all other criteria are irrelevant.

We can normally find ways to handle the first and second limitations. For example, the second limitation can be regulated with the ideas of risk management described in Section 4.5.

The truly innovative products, those that change the market, are the real concern. Even the vocabulary used to describe these products is different. They are "show stoppers," "game changers," or "step-out technologies." They are "out of the box," that is, conceived beyond current thinking. They are hard to find and hard to recognize.

As one vivid example, imagine that you are William Caxton in 1476. You are trying to decide whether it makes sense to print Chaucer's "Canterbury Tales,"

TABLE 3.4–I   Concept-Screening Matrix for Printing Chaucer's Canterbury Tales

Selection Criteria	Weighting Factor	Illuminated Manuscript	Printed Chaucer
Quality	0.4	5	1
Cost	0.4	5	6
Quantity	0.2	5	8
Total Score		5	4.4

Note: This matrix could be one developed by William Caxton, in 1476.

or continue to produce illuminated manuscripts. You might develop the concept-scoring matrix shown in Table 3.4–1. Your three scoring criteria are the quality of the finished product, its cost to produce, and the quantity that you can make. The illuminated manuscript is the benchmark and so gets a score of 5 for each criterion. The illuminated manuscript is of much higher quality than a smudged and crooked printed version of Chaucer, so the printed edition gets a low score for quality. The cost of the printed Chaucer is less, but that is not a big advantage: the monks, those postgraduate students of the Middle Ages, work for nothing. What would they do if they were not copying manuscripts? Finally, you can make more printed Chaucers. But the market is limited to the literate, mostly the monks. Thus the quantity you can make is not important. Thus, as Table 3.4–1 shows, it does not make sense to print books.

With historical hindsight, we know this conclusion is wrong: it does make sense to print books. The use of moveable type in printing was an enormous technical innovation. However, we should admit that recognizing such advances at the time will always be hard. In this sense, we should remember that when he first printed Chaucer, William Caxton was not a printer. He was a wool merchant: perhaps, he was looking for an outlet for excess sheepskins, which are of course the feedstock for parchment. He may have begun printing Chaucer to use up extra sheepskins. With this cautionary example in mind, we turn to a harder, technical example.

EXAMPLE 3.4–I   HOME OXYGEN SUPPLY

Those with lung disorders, including emphysema, can sometimes benefit from breathing air enriched with oxygen. This oxygen is presently supplied as cylinders of nearly pure oxygen, regularly delivered, just as those who live in remote areas may have regular deliveries of propane for use in cooking. Oxygen delivered in cylinders works well, but can be expensive. Shifting the cylinders around in the house can be difficult, especially if the user is older.

We want to find an alternative to gas cylinders to provide home oxygen. Our idea generation and sorting has led to two reasonable alternatives: membrane separation and pressure swing adsorption (PSA). The membrane separation uses selective hollow fibers in a module like a shell and tube heat exchanger, except

TABLE 3.4–2   Concept-Screening Matrix for Home Oxygen Supply

Selection Criteria	Weighting Factor	Gas Cylinders	Hollow-Fiber Membranes	PSA
Convenience	0.4	5	8	8
Noise	0.3	5	4	2
Cost	0.3	5	7	7
Total Score		5	6.5	5.9

Note: Both membranes and PSA score better than cylinders delivered containing oxygen. However, no single process stands out compellingly.

with tubes less than 1 mm in diameter. It requires a pump to compress room air and force it across the fibers. This permeate air will contain perhaps 30% oxygen.

The PSA unit uses an adsorbant, which can be either oxygen or nitrogen selective. The adsorbant is often a zeolite. If the adsorbant is oxygen selective, then air at high pressure is forced through until the adsorbant is saturated. Then the flow is stopped and the pressure is released. The air coming out of the bed is enriched with oxygen. This system also requires a pump, as well as some valving.

Choose key factors and evaluate which of these ideas is better.

SOLUTION

In this simple case, the core team decided that there are three key factors: convenience, noise, and cost. The three factors are of roughly equal importance, with convenience marginally the more important. Noise may be unimportant to a geriatric patient who is deaf but it may be important to anyone who lives with the patient. Cost could be important if the patient pays for the system himself. However, the core team could decide that these costs will largely be borne by insurance or by health maintenance organizations, and that cost differences are not so important.

On this basis, we can prepare a concept-screening matrix like that shown in Table 3.4–2. This matrix shows the three selection criteria on the left-hand side of the table. The weighting factors for these criteria are shown in the second column, with convenience given a slightly greater importance than either noise or cost. The remaining three columns in the table give the scores of the three alternatives. Note that because it is the benchmark, the gas cylinder is always given the arbitrary score of 5. Note also that the hollow fiber membranes have the top score, followed by the PSA. However, we may find it harder to make a membrane with the desired properties than to locate an effective zeolite adsorbent.

EXAMPLE 3.4–2   HIGH-LEVEL RADIOACTIVE WASTE

This complex example is unusual in that the product is actually a chemical plant. The manufacturer of the product will be one of a handful of aerospace companies; the customers will be the US Department of Energy, British Nuclear Fuels Limited, and other similar agencies. The example is included because it dramatically illustrates how the ideas of product design tend to precede those of process design, as discussed in Section 1.4.

The technical problem is as follows. In the manufacture of atomic weapons, a significant number of by-products are made. Many of these by-products are dangerously radioactive isotopes; most of these are actinides which precipitate in basic solution. These precipitated isotopes are then separated and vitrified, that is, made into glass. The glass is sufficiently radioactive that it can boil water. It must be safely stored for thousands of years.

However, one highly radioactive isotope of cesium, ^{137}Cs, is not precipitated in base but remains dissolved in aqueous solutions. Millions of gallons of this aqueous solution are stored in aging tanks in the locations where the atomic weapons were manufactured. If the tanks leak because of aging or earthquakes, the escaping cesium would spell disaster.

Not surprisingly, there has been a recent major effort to develop a means to make the ^{137}Cs less dangerous. Some of the 180 serious ideas suggested for this are shown in Table 3.1–3. These ideas are sorted as shown in Table 3.4–3. The organization has an unsurprising form: the first heading (I) deals with improvements to the existing process; the next four (II to V) are essentially separation processes organized as unit operations; and the last heading (VI) centers on innovations. The innovations include technically feasible ideas that are certainly politically unacceptable. In particular, idea 23 suggests setting off nuclear weapons inside the existing storage tanks to use the heat to vitrify the entire contents of the tank *in situ*. It could work, and it would be inexpensive because the nuclear weapons are in inventory; but it is not politically feasible or morally justified.

Develop a concept-screening matrix which can choose among the ideas in Table 3.4–3 to find a small number for further development.

SOLUTION

Because this problem is so complex, we will offer only a partial solution here. Details of the decision are only partially available publicly, and the final choice of a treatment has not yet been made (in 2001). Still, the problem is a superb example of the power of the methods suggested in this section.

We begin by describing our benchmark, the existing process, and by choosing the screening criteria. The existing process hinges on the precipitation of the cesium cation with the tetraphenylborate anion:

$$Cs^+ + B(C_6H_5)_4^- \rightarrow CsB(C_6H_5)_4.$$

This anion also precipitates potassium ions, present in nonradioactive form at much higher concentrations than cesium. However, the sodium ion, which is the chief cation present, does not precipitate. Thus we can precipitate the radioactive cesium by adding saturated solutions of sodium tetraphenylborate. The resulting process separates and concentrates the cesium about 20,000 times.

In the existing precipitation process, the cesium precipitate is made in large quantities and then stored. It has to be slowly added to the other radioactive precipitates, and eventually vitrified into glass. Unfortunately, the cesium precipi-

---

TABLE 3.4–3 An Outline of Ideas for High-Level Waste

I. Improve Current Process
   A. Increase storage capacity (14, 17)
   B. Stabilize precipitate (4, 6, 36, 39)
   C. Legalize – more benzene release (41)
   D. Vitrify without separation (38, 49)

II. Separate by Precipitation (45)
   A. Process alternatives
      a. Make fast (10)
      b. Flocculate (32)
   B. Stabilize current precipitate (15)
      a. With catalyst poison (6)
      b. By chilling tanks (18)
   C. New selective precipitate (12, 42, 46)

III. Separate by Adsorption (44)
   A. Process alternatives
      a. Simulated moving bed (21)
      b. Fluidized bed (29, 47)
   B. Regenerable ion exchange (13, 25)
      a. Crown ethers (35)
      b. $MnFe(CN)_6$ (31)
      c. Electrically switched ion exchange (27, 34)
   C. Nonregenerable ion exchange
      a. On glass
      b. On crystalline silicon titanate (2, 24)

IV. Separate by Extraction
   A. Process – hollow-fiber membranes (16)
   B. New chemistry
      a. Caustic with zeolite or crown ether (3, 5)
      b. Acidic (26)

V. Less Conventional Separations
   A. Fractional crystallization (8, 33, 43)
   B. Decantation (40)
   C. Electrochemical methods (7, 11, 37, 48)
   D. Microbes (23)

VI. Stabilize Without Separation
   A. Into bedrock (1, 22)
   B. As concrete grout (9, 20, 30)

Note: The numbers refer to the specific ideas listed in Table 3.1–3.

tate is not stable: it decomposes to produce soluble cesium and benzene. Because benzene is volatile, the tanks where the precipitate is stored can have inflammable benzene vapor collecting in the headspace above the precipitate. As a result, this headspace is continually flushed with nitrogen. Although storing the unstable precipitate may look foolish now, the precipitate was not expected to be unstable.

TABLE 3.4–4   Sample Scores for Four of the Processes for Treating Radioactive Waste Containing ^{137}Cs

Selection Criterion	Weighting Factor	Precipitation and Storage Benchmark	Precipitation and Treatment (II.A.a)	Crown Ether Ion Exchange (III.B.a)	Crown Ether Extraction (IV.B.a)	Concrete Formation (VI.B)
Mature science	0.4	5	9	3	2	6
Reliable engineering	0.4	5	7	6	9	8
Safety	0.1	5	5	4	8	10
Public response	0.1	5	5	5	5	1
Total score		5	7.4	4.5	5.7	6.7

Note: The screening procedure described and the scores are a simplification of the actual method used.

It is unstable because the waste contains parts per billion of palladium, which catalyzes this decomposition. The active role of this palladium was unexpected.

The screening criteria for this waste treatment center on finding a process that works now. This means that the two most important criteria are known science and reliable engineering. Each of these was given a weighting factor of 0.4. Two other important criteria are safety and the public response, each of which was given a weighting factor of 0.1. Note that cost does not appear in these criteria, an implicit recognition that all of these choices will be expensive. Note also the relatively low weighting factor for public response. This does not imply that public response is unimportant, or that the public may not veto any one of the choices. At this point, however, those involved – the core team – felt that choosing a process that was scientifically and technically reliable was paramount.

Each of the processes in Table 3.4–3 is then scored by using these criteria. Typical scores for four of these processes are shown in Table 3.4–4. These four are all reasonable alternatives: each could work well.

The processes in Table 3.4–4 merit more detailed discussion. "Precipitation and fast treatment" is closest to the existing process. It begins by precipitating the cesium tetraphenylborate, but under more controlled conditions than those used at present. Once the precipitate is formed, it is separated and forwarded to vitrification before significant decomposition can occur. For this option, the science and engineering are in good shape, and the safety and public response should be similar to that for the existing process.

The second and third options both replace the tetraphenylborate precipitation with the new chemistry of macrocyclic ethers, exemplified by the compound dibenzo-18-crown-6:

Compounds like this species can selectively complex specific cations like cesium. Some specific materials suitable for both ion exchange and liquid-liquid extraction are known. However, only small amounts (perhaps 100 g) have been made. Commercial suppliers are either nonexistent or underfunded academic spin-offs. This means that both these options have low scores on scientific maturity. Nonetheless, an alternative like this was attractively developed by a laboratory at one of the sites where atomic weapons had been produced.

The final "concrete formation" option is the most risky. In this option, we would not try to separate the cesium at all, but simply turn the entire contents of the storage tank into a concrete-like solid, called "grout." While the science for such a high-salt grout needs a little work, the engineering is straightforward. The process is safe: because we do not concentrate the cesium, the waste remains in the less dangerous "low-level" form.

The risk is the public's response. Instead of concentrating the waste and shipping it to some distant desert storage, the waste is simply immobilized and left where it is. Local citizens and their elected officials will not like that, even though they have been happy to have the jobs that making atomic weapons has supplied. Although making grout is legal within the letter of existing laws, the public's response may engender so much litigation that the solution of the problem of radioactive waste is delayed and the danger of accidental release is increased. Still, this alternative is a strong contender. In more general terms, you can now see why we wanted such a complex example: it really does show how different alternatives must be carefully weighed.

3.5   Conclusions and the Second Gate

This chapter describes the generation and screening of ideas, which is the second step in our four-step template of product design. Before product ideas are developed, we identify product needs and quantify these needs with product specifications. Identifying needs and developing specifications was the subject of the previous chapter. After these ideas are screened, we must select among the best choices and manufacture the products. Selection and manufacturing are described in later chapters.

We normally welcome product ideas from every possible source. These sources center on the core team but also include customers, competitors, and consultants. We will depend most heavily on "brainstorming" to generate without criticism a broad selection of concepts. A useful target is around 100 such ideas.

Because we will rarely have the resources to evaluate completely all of these often fragmented ideas, we need to choose the best for further study. Such choices can be made in two stages. First, we organize the ideas, removing redundancy and pruning folly. This will normally give us around twenty remaining candidates. We will then use concept-screening methods to judge the advantages and disadvantages of these ideas. This will normally reduce the number to five or fewer. Cutting this number further hinges heavily on engineering and chemistry, as described in the next chapter.

This is also the point of the second management review, the second "gate" through which a successful product development must pass. Again, the core team will make a presentation to the same senior management group as for the first gate. Again, this presentation will include both oral and written components. As at the first gate after "needs," management will decide whether or not to continue work.

Interestingly, business studies seem to show that management again tends to be too supportive at this stage. The management team may be charmed by suggested innovations and excited by product improvements. When the management team have nontechnical backgrounds such as law or marketing, they may require help in making rational decisions if a strong chemical component is involved. As a result, the core team must be especially careful to be objective. The team must make sure that management understands not only the potential rewards, but also the risks. After this second gate, product development gets more expensive as we select the product that we hope to manufacture.

FURTHER READING

Appleby, A. J. and Foulkes, F. R. (1989). *Fuel Cell Handbook*. Van Nostrand Reinhold, New York, ISBN 00442319266.

Cox, P. A. and Balick, M. J. (1994). The Ethnobotanical Approach to Drug Discovery. *Sci. Am.* **270**, 82–87.

Engstrom, J. R. and Weinberg, W. H. (2000). Combinational Materials Science. *AIChE J.* **46**, 2–5.

Farnsworth, N. R. (1990). The Role of Ethnopharmacology in Drug Development. *Ciba Found. Symp.* **154**, 2–21.

Linden, D. (ed.) (1995). *Handbook of Batteries*, 2nd ed. McGraw-Hill, New York, ISBN 0070379211.

Lowe, G. (1995). Combinatorial Chemistry. *Chem. Soc. Rev.* **24**, 309–317.

Murray, R. L. (1993). *Nuclear Energy: An Introduction to the Concepts, Systems and Applications of Nuclear Processes*, 4th ed. Butterworth-Henemann, Oxford, ISBN 0750628952.

Reddington, E., Sapienza, A., Gurau, B., Viswanathan, R., Sarangapani, S., Smotkin, E. S., and Mallouk, T. E. (1998). Combinatorial Electrochemistry: A Highly Parallel, Optical Screening Method for Discovery of Better Electrocatalysts. *Science* **280**, 1735–1737.

4

Selection

As explained earlier, we expect that product design will take place in four, roughly sequential steps. First, we will identify needs; then, we generate ideas to fill these needs; third, we select the best ideas; and last, we consider manufacturing. So far, we have discussed identifying needs, generating ideas to fill these needs, and choosing a shorter list for further study.

We now want to select the best ideas for further development. In some cases, we will have only one or two clear choices. In most cases, we will want to select five or fewer, simply because the amount of work required for further development is so substantial. In selecting these few products, we can identify two separate situations. In the first, we can compare products by using only chemical and engineering criteria. In the second, we must compare products not only on a technical basis, but also by using less exact criteria, like "comfort" and "safety." How we proceed in product selection depends on which of these two situations we encounter.

In the first situation involving only chemical and engineering criteria, we already have the tools for selection from our technical training. In particular, we have a background in thermodynamics and so can calculate any chemical equilibria or heats of reaction suggested in our product development. We have a background in kinetics, especially a knowledge of reaction rate constants, including how these change with temperature. Those trained in chemical engineering will also have a knowledge of heat and mass transfer.

Because of this background, we should be able to make the approximate calculations necessary for product selection. However, in our own experience we find that some aspects that are important in product selection are not stressed in conventional courses. We review these topics in this chapter. In Section 4.1, we discuss neglected thermodynamics topics. In Section 4.2, we review some kinetics topics.

The second situation, in which selection includes both technical and less exact criteria, is more difficult than the first, which is based firmly in chemistry and engineering. Often these less exact criteria will include consumer reactions and public opinion. These criteria may change from one country to another, or evolve over time. In the vernacular, we may describe this second situation as selecting between "apples and oranges." In the first situation, we would be making the easier comparison between "apples and apples."

In the final sections of the chapter, we outline some ways to select between dissimilar products, that is, among "apples and oranges." In Section 4.3, we again use the weighting factors introduced in Section 3.4 to make judgments combining subjective and objective considerations. In other words, we will try to balance factors such as "comfortable" with factors such as "heat recovery." In Section 4.4, we explore the effect of risk. Our goal is to decide between an effective but expensive idea and a potentially superior idea that may not work. At the end of this effort, we should be prepared to select our new product, and be ready to move to the final manufacturing step.

4.1   Selection Using Thermodynamics

The selection of the best potential product out of a short list of carefully culled ideas is easiest when the proposed new products are modifications of existing products. These modifications most commonly involve ingredient substitutions or improvements in performance. They are often based in thermodynamics.

INGREDIENT SUBSTITUTIONS

Changes in a product's chemical ingredients usually seek to duplicate its current properties. The most common effort is the search for less volatile, less toxic solvents. For example, methylene chloride (CH_2Cl_2), one of the most useful solvents used for fine chemical manufacture, is a carcinogen. That laboratory mainstay, acetone (CH_3COCH_3), is more toxic than methanol (CH_3OH). We want to replace solvents believed to be dangerous with a benign solvent that is nonvolatile, nontoxic, and cheap. We want to equal product performance, but with additional benefits, such as safety and cheapness.

The best route to discovering new solvents is by experiments. Although we will almost always need these, we can also benefit from some kind of guide that will let us choose good candidates. This guide should be simple and quick, rather than accurate but tedious.

The best such guide uses solubility parameters. This guide, originally suggested by Hildebrand and effectively extended to polymers by Hansen, assumes that all solutions are nonideal, and are described by a relation of the following form:

$$\mu_2 = \mu_2^0 + RT \ln x_2 + \omega x_1^2, \tag{4.1--1}$$

where μ_2 is the chemical potential of the product solute; μ_2^0 is its value in a reference state (pure 2 at the specified T and p), the "standard state"; ω is an activity parameter with the dimensions of energy per mole; and x_1 and x_2 are the mole fractions of the solvent and the product, respectively. The logarithmic term in this equation represents the free energy change of ideal mixing, and is related to entropy changes. The term with ω includes any heat of mixing. This simplest nonideal relation is known as the Margules equation.

The parameter ω is the key to selecting alternative solvents. It can be estimated by

$$\omega = \underline{V}_2(\delta_1 - \delta_2)^2, \tag{4.1-2}$$

where $\underline{V}_2$ is the molar volume of the product solute, and the δ are the so-called solubility parameters. Because ω has dimensions of energy per mole and $\underline{V}_2$ is a molar volume, the δ are given in the improbable units of $[\text{energy/volume}]^{1/2}$. For an ideal solution, ω is zero, and the solute and solvent are miscible in all proportions. For a nonideal solution, the solubility parameters will differ. The more different they are, the less miscible the solute and the solvent will be. Note that ω cannot be negative in this simple model and so the nonideality is not allowed to favor mixing.

In almost all cases, we will not know the solubility parameter δ_2 for the product solute. We do have tables of the solubility parameters of common solvents, a selection of which is given in Table 4.1–1. To find a solvent giving the same properties as that we are using, we simply seek a solvent with δ_1 equal to that of the current solvent. For example, if we are currently using chloroform with δ_1 of 9.2 $(\text{cal/cm}^3)^{1/2}$, we could try substituting benzene, which has the same δ_1 value. We will use this strategy in an example later in this section.

SUBSTITUTIONS IN CONSUMER PRODUCTS

At this point, we interrupt outlining the methods of ingredient substitution to touch on some special characteristics of consumer products. There are two important ones. The first is that isomerically pure chemicals are the same no matter whether their origin is the farmer's field or the laboratory. We recognize that this may be anathema to those interested in "natural foods" or "natural ingredients," but it is true. If a particular chemical isomer is pure, then a sample obtained from vegetable sources is indistinguishable from a similar sample made in a laboratory.

Because of this equivalence, we should be able to substitute a "natural" chemical for a "synthetic" chemical without any difficulty whatsoever. The reverse is equally true. If we do have trouble, it is almost certainly because one of the chemicals is impure. In the vast majority of cases, the impure chemical is the "natural" one. Now this impure chemical can be superior; for example, it might contain an emulsifier or a surfactant that facilitates its mixing. Still, if there is a problem in making a natural versus a synthetic substitution, first look for the problem in an impure natural material.

We are not arguing that natural ingredients are less desirable than synthetic ingredients. In chemical terms, we repeat that pure materials are equivalent. However, there may be major marketing advantages in using natural ingredients, for they are often perceived as superior, and command premium prices. Curiously, any chemicals made microbiologically can legally be identified as natural. Thus, impure pepper flavoring made from genetically modified organisms that would be completely incapable of survival outside the laboratory may be labeled as natural, whereas pure flavoring made in aseptic reactors cannot be. Life is strange.

TABLE 4.1–1   Solubility Parameters

Formula	Substance	$\underline{V}$ (cm³/mol)	δ (cal$^{1/2}$ cm$^{-3/2}$)
Halogenated solvents			
C_6F_{14}	perfluoro-n-hexane	205	5.9
C_7F_{16}	perfluoro-n-heptane	226	6.0
C_6F_{12}	perfluorocyclohexane	170	6.1
$(C_4F_9)_3N$	perfluoro tributylamine	360	5.9
$C_2Cl_3F_3$	1,1,2-trichloro,1,2,2-trifluoroethane	120	7.1
CH_2Cl_2	methylene chloride	64	9.8
$CHCl_3$	chloroform	81	9.2
CCl_4	carbon tetrachloride	97	8.6
$CHBr_3$	bromoform	88	10.5
CH_3I	methyl iodide	63	9.9
CH_2I_2	methylene iodide	81	11.8
C_2H_5Cl	ethyl chloride	74	8.3
C_2H_5Br	ethyl bromide	75	8.9
C_2H_5I	ethyl iodide	81	9.4
$C_2H_4Cl_2$	1,2 dichloroethane (ethylene chloride)	79	9.9
$C_2H_4Cl_2$	1,1 dichloroethane (ethylidene chloride)	85	9.1
$C_2H_4Br_2$	1,2 dibromoethane	90	10.2
$C_2H_3Cl_3$	1,1,1 trichloroethane	100	8.5
Aliphatic hydrocarbons			
C_5H_{12}	n-pentane	116	7.1
	2-methyl-butane (isopentane)	117	6.8
	2,2-dimethyl propane (neopentane)	122	6.2
C_7H_{14}	n-hexane	132	7.3
C_7H_{16}	n-heptane	148	7.4
C_8H_{18}	n-octane	164	7.5
	2,2,4-trimethylpentane	166	6.9
$C_{16}H_{34}$	n-hexadecane	294	8.0
C_6H_{12}	cylohexane	109	8.2
C_7H_{14}	methylcyclohexane	128	7.8
C_6H_{12}	1-hexene	126	7.3
C_8H_{16}	1-octene	158	7.6
C_6H_{10}	1,5-hexadiene	118	7.7
Aromatic hydrocarbons			
C_6H_6	benzene	89	9.2
C_7H_8	toluene	107	8.9
C_8H_{10}	ethylbenzene	123	9.9
	o-xylene	121	9.0
	m-xylene	123	8.8
	p-xylene	124	8.8
C_8H_8	styrene	116	9.3
$C_{10}H_8$	naphthalene	123	9.9
Inorganics			
Br_2	bromine	51	11.5
I_2	iodine	59	14.1
S_8	sulfur	135	12.4
P_4	phosphorus	70	13.1
CCl_4	carbon tetrachloride	97	8.6
$SiCl_4$	silicon tetrachloride	115	7.6
$SnCl_4$	stannic chloride	118	8.7
WF_6	tungsten hexafluoride	88	8.0
$Si(CH_3)_4$	silicon tetramethyl	136	6.2

List reproduced from Hildebrand, Prausnitz, and Scott.

In addition to this natural versus synthetic issue, a second characteristic of consumer products that affects ingredient substitutions concerns the evaluation of consumer attributes. If we are going to make an ingredient substitution, we need a way of assessing success or failure. As an example, we consider three attributes of skin creams: "thickness," "smoothness," and "creaminess." As discussed in Section 2.4, "thickness" is proportional to the square root of instrumentally measured Newtonian viscosity. From other studies, we know that "smoothness" is related to the coefficient of friction. More specifically, "thickness" is proportional to the force of viscous drag; but "smoothness" is inversely proportional to the frictional force during contact lubrication. Interestingly, "creaminess" seems close to the geometric average of "thickness" and "smoothness." It is not an independent attribute.

In a case like this, we can easily substitute ingredients because we know which consumer attributes are related to which physical properties. An example illustrating such substitutions is given later in this section. Unfortunately, we usually do not know these relationships, so that ingredient substitution becomes empirical. This relationship between consumer attributes and physical properties, which currently falls between social science and engineering, would greatly benefit from more study.

INGREDIENT IMPROVEMENTS

We will also frequently want to improve products by using ingredients that are superior. It is often useful for ingredients to have properties that are strong functions of temperature or pH, because this allows the product's activity to be triggered. Products whose properties change dramatically with temperature are the more common case. An excellent example is a class of water soluble absorbents, like monoethanol amines, which are used to scrub acid gases. These amines react rapidly with gases such as carbon dioxide to form water soluble products, in this case carbonates. The carbonates are easily destroyed by gentle heating, thus reversing the reactions. However, as energy costs rise, there is a major interest in amines whose reactions are not only fast, but whose equilibria with acid gases change radically with temperature.

To seek such amines, we again return to simple thermodynamic ideas. To focus these ideas, we consider the reaction

$$H_2S(g) + RNH_2(aq) \rightleftharpoons RNH_3^+(aq) + HS^-(aq). \tag{4.1-3}$$

Thus,

$$[RNH_2^+][HS^-] = K[RNH_2][H_2S], \tag{4.1-4}$$

where the quantities in square brackets are concentrations and K is the equilibrium constant for this reaction.

We want K to vary strongly with temperature. To seek this variation, we remember that

$$K = e^{-\Delta \underline{G}/RT} = e^{-\Delta \underline{H}/RT + \Delta \underline{S}/R}, \tag{4.1-5}$$

where $\Delta \underline{G}$, $\Delta \underline{H}$, and $\Delta \underline{S}$ are the free energy, enthalpy, and entropy changes of this reaction. Thus, we seek amines that show large enthalpies of reaction when combined with H_2S.

As a case illustrating variation of a property with pH, we consider drug purification. Many drugs have either carboxylic acid ($-COOH$) or amine ($-NH_2$) groups that can be used to facilitate their purification by liquid-liquid extraction. As an example, we imagine an antibiotic like penicillin, which is a carboxylic acid. We expect that at equilibrium, the concentrations in an organic solvent and in water are at equilibrium:

$$[\text{drug in organic}] = K[\text{drug in water}],$$

$$[RCOOH]_{org} = K[\overline{RCOOH}]_{H_2O}, \tag{4.1-6}$$

where K is again an equilibrium constant. The drug in organic solution will normally be protonated, and hence in only one form. However, the drug in aqueous solution may either be protonated or ionized.

The ionization of the drug in aqueous solution means that the equilibrium constant K can be a strong function of pH. To see why, we first note that the concentration $\overline{RCOOH}$ actually includes different forms:

$$[\overline{RCOOH}]_{H_2O} = [RCOOH]_{H_2O} + [RCOO^-]_{H_2O}. \tag{4.1-7}$$

These forms are in equilibrium:

$$[H^+]_{H_2O}[RCOO^-]_{H_2O} = K_a[RCOOH]_{H_2O} \tag{4.1-8}$$

where K_a is the dissociation constant for this acidic drug. The true thermodynamic equilibrium across the organic-water interface is not that in Equation 4.1–6, but

$$[RCOOH]_{org} = K'[RCOOH]_{H_2O}. \tag{4.1-9}$$

Combining Equations 4.1–7 to 4.1–9, we find that

$$[RCOOH]_{org} = \left(\frac{K'[H^+]_{H_2O}}{K_a + [H]_{H_2O}} \right) [\overline{RCOOH}]_{H_2O}. \tag{4.1-10}$$

Thus

$$K = \frac{K'[H^+]_{H_2O}}{K_a + [H^+]_{H_2O}}. \tag{4.1-11}$$

The apparent equilibrium constant K in Equation 4.1–6 varies dramatically with acid concentration. This variation, explored further in an example below, is useful in drug purification by extraction.

EXAMPLE 4.1–1   A BETTER SKIN LOTION

Our employer currently manufactures a variety of skin care products. Market research shows a need for a "thinner," "creamier" skin lotion. Our idea is to develop such a lotion that is twice as "thin" (i.e., half as "thick") but also twice as "creamy." How should we seek new ingredients?

SOLUTION

To make the lotion half as thick, we can simply reduce its viscosity. If the lotion is Newtonian, we should reduce its viscosity four times. However, to double the lotion's "creaminess," we must increase its "smoothness" four times, because past experiments show that

$$\text{"creaminess"} = [(\text{"thickness"}) \times (\text{"smoothness"})]^{1/2}$$

This increased "smoothness" requires decreasing the coefficient of friction by eight times. This may be hard to do.

EXAMPLE 4.1–2 A POLLUTION PREVENTING INK

For the purposes of this example, a lithographic ink can be idealized as containing only four components: a pigment, an oil, a resin, and a solvent. The pigment, frequently colloidal carbon, is important to the ink, but not a key in pollution. The oil, a mixture of natural products such as castor and linseed oils, typically contains fatty acids with multiple double bonds, such as linoleic acid and linolenic acid. These double bonds crosslink in the presence of oxygen, making the ink permanent. The resin is a low molecular weight, highly polydisperse condensation polymer, made for this purpose. The solvent, frequently methylene chloride (CH_2Cl_2), is used to adjust the ink's rheology to give good printing.

We want to replace methylene chloride with a less dangerous solvent. What should we choose?

SOLUTION

We must seek solvents whose solubility parameters are close to that of methylene chloride. From Table 4.1–1, we find that the solubility parameter of methylene chloride is 9.8 $(\text{cal/cm}^3)^{1/2}$. We then look for inexpensive solvents with similar solubility parameters. Benzene, toluene, and naphthalene are three possible choices. All are carcinogens, though felt to be less dangerous than methylene chloride. Benzene is excessively volatile, and naphthalene is solid at the temperatures normally used for printing. Thus we choose toluene as a substitute for methylene chloride and start experiments. This is a conservative choice: the toluene is still toxic and will generate emissions, but the modified ink should work well.

EXAMPLE 4.1–3 ANTIBIOTIC PURIFICATION

We are trying to modify an existing purification to use with a new acid antibiotic whose pK_a is 4.52. We want to alter the pH and hence alter the distribution of the antibiotic between water and butyl acetate. What pH range should we use?

SOLUTION

To begin, we should return to the definition of the pK_a, which is

$$pK_a = -\log_{10} K_a = \log_{10} \frac{[\text{H}^+]_{\text{H}_2\text{O}}[\text{RCOO}^-]_{\text{H}_2\text{O}}}{[\text{RCOOH}]_{\text{H}_2\text{O}}},$$

where the various concentrations are defined by Equation 4.1–8. The definition of the pH is

$$pH = -\log_{10}[H^+].$$

Thus from Equation 4.1–11, we see

$$\frac{K}{K'} = \frac{1}{1 + 10^{pH - pK_a}}.$$

When the pH is much less than 4.52, K approaches K'. When the pH is much greater than 4.52, K becomes small. By adjusting the pH, we can change K and purify the drug.

4.2   Selection Using Kinetics

If thermodynamics is the science of the possible, then kinetics is the science of how fast and hence how expensive. For chemical products, we must normally consider three kinds of kinetics. We must obviously consider the rates of the chemical reactions by which raw materials become the desired products. Reaction rates normally can not be predicted a priori, so this implies experimental data. However, even when data are missing, we can make estimates of the maximum rates possible. The other two forms of kinetics are mass and heat transfer rates. Here, we are in better shape, because we can estimate heat and mass transfer coefficients. In the following paragraphs, we briefly review each of these kinetics topics to supply guidelines for product selection.

CHEMICAL KINETICS

Chemical reaction rates must in almost all cases be determined by experiment. Books describing these rates all describe the more common reaction mechanisms. For a first-order reaction, the rate per volume r is

$$r = kc_1, \tag{4.2-1}$$

where c_1 is the concentration of reactant species. "1" and k is the reaction rate constant, with units of reciprocal time. For a second-order reaction, the rate is

$$r = (k'c_2)c_1, \tag{4.2-2}$$

where c_2 is the concentration of a second reagent, and k' is a second-order reaction rate constant, now with dimensions of volume per mole per time. Note that when c_2 is present in excess, c_1 changes by a much greater percentage than c_2. Then the quantity in parentheses in Equation 4.2–2 is nearly constant, a pseudo-first-order rate constant. This is one reason why so many reactions with more than one reagent can appear to be first order.

Zero-order reactions are also surprisingly common, for which the rate r is

$$r = k'', \tag{4.2-3}$$

where k'' has the dimensions of moles per volume per time. At first blush, it may seem amazing that the reaction rate will not change if the amount of reagent is increased. In fact, zero-order reactions are a frequent result of chemical catalysis, for each molecule of catalyst can only react so fast. This is explored further in Section 5.5.

The important point here is that no reaction rate constant – k, k', or k'' – can be found without experiment. Without experiments, we are really guessing, with substantial risk. This uncertainty may cause us considerable difficulty in product selection. However, things are not as bleak as this experimental demand may seem, because we can predict the maximum reaction rate. This is because as the chemical steps in the reaction become fast, the reaction rate is controlled by mass transfer. Mass transfer is much more easily estimated. If this maximum rate for a specific process is slower than we need, the process should not be selected.

The prediction of the maximum rate is easier for heterogeneous reactions. For simplicity, we will discuss only first order, irreversible cases. For example, imagine we are interested in the combustion of solid particles in air. The apparent rate constant k of such a reaction is

$$\frac{1}{k} = \frac{1}{k_D a} + \frac{1}{k_{\text{surface}}}, \tag{4.2-4}$$

where k_D is a mass transfer coefficient, a is the particle area per volume, and k_{surface} describes the kinetics of the reaction on the particle surface. If this surface reaction is fast, k_{surface} is very large, and the process will be controlled by mass transfer. Such mass transfer can be estimated simply by assuming k_D is about 10^{-3} cm/sec in a liquid or 1 cm/sec in a gas. The rate constant k is then known if the area per volume a is known. We will return to this estimate again later in this section.

Many reactions with microorganisms, catalyst particles, or emulsion droplets show these same characteristics. Correlations of mass transfer coefficients exist for these systems, but they really aren't the point. If we can guess the area per volume, we can guess the maximum apparent rate constant k, and hence have a more quantitative basis for selecting between alternatives.

If the reaction is homogeneous, the estimates of the maximum rate are harder. To see why, we consider the simplest limit of instantaneous, irreversible reactions. If the system is initially well mixed and the reaction chemistry is instantaneous, then

$$k = 4\pi(D_1 + D_2)\sigma \tilde{N} c_2, \tag{4.2-5}$$

where D_1 and D_2 are the diffusion coefficients of the limiting and excess reagents, respectively; σ is a collision diameter, perhaps 5 Å; $\tilde{N}$ is Avogadro's number; and c_2 is the concentration of the excess reagent. Unfortunately, this often-quoted result is not useful in practice, because systems which are highly reactive don't start out well mixed.

Instead, many fast homogeneous reactions have their reaction rates determined by the mixing. In general, the rate constant for this mixing is

$$k = [4(D_1 + D_2)/l^2], \tag{4.2-6}$$

where l is the average eddy size of the mixing. In rapidly stirred liquids, l is about 30 μm and the D_i are 10^{-5} cm^2/sec, so k is about 10^4 sec^{-1}. If there are no experiments, this can provide a quick guess of maximum possible reaction speed. We must repeat that whenever available, experiments are better than estimates.

HEAT AND MASS TRANSFER COEFFICIENTS

Finally, we need to turn to heat and mass transfer rates. Because these are physical processes, not chemical changes, the rates vary much less widely. They can be more easily estimated, at least to an order of magnitude.

Mass transfer coefficients vary least. For liquids, the most common value of mass transfer coefficient is that given above, about 10^{-3} cm/sec. One easy way to think about the physical significance of this estimate uses the film theory of mass transfer, for which

$$k = D/\delta, \tag{4.2-7}$$

where D is the diffusion coefficient and δ is a "boundary layer" or a "film thickness," roughly the distance over which the concentration changes. Because liquid diffusion coefficients cluster around 10^{-5} cm^2/sec, this implies a boundary layer of 100 μm, that is, 0.01 cm.

For gases, values of mass transfer coefficients fall in a wider range, but tend to be around 1 cm/sec. Because diffusion coefficients are about 0.1 cm^2/sec, this implies a boundary layer of perhaps 0.1 cm. The wider variation of coefficients in the gas is rarely a problem because most mass transfer is limited by the liquid.

Heat transfer coefficients cluster near these same values, but this clustering is obscured by the units used. To see why this is so, we turn to that classic tool of chemical engineering, the Chilton-Colburn analogy:

$$\frac{k}{v}\left(\frac{v}{D}\right)^{2/3} = \frac{h/\rho\hat{C}_p}{v}\left(\frac{v}{\alpha}\right)^{2/3} = \frac{f}{2}, \tag{4.2-8}$$

where k and h are the mass and heat transfer coefficients, respectively; ρ is the density; $\hat{C}_p$ is the specific heat capacity; v is the velocity; f is the friction factor; and D, α, and v are the diffusivities of mass, energy, and momentum. The quantity α is also called the thermal diffusivity, and v is also called the kinematic viscosity. The important point is that these quantities are physical properties, listed in handbooks.

We can use Equation 4.2–8 to estimate heat transfer coefficients. For gases, α and D are both around 1 cm^2/sec, so

$$k = h/\rho\hat{C}_p. \tag{4.2-9}$$

This result is sometimes called the Reynolds analogy. Because k in gases is around 1 cm/sec, $(h/\rho\hat{C}_p)$ is also around 1 cm/sec. For liquids, α is around 10^{-2} cm^2/sec

TABLE 4.2–I   **Rough Estimates of Heat and Mass Transfer Coefficients**

Situation	k (cm/sec)	$h/\rho\hat{C}_p$ (cm/sec)	h (W/m^2°K)
Flowing gases	1	1	3
Flowing water	0.001	0.1	5000
Flowing organics	0.001	0.1	1000
Condensing steam or boiling water	—	—	2000

Note: The values given are useful only in the order of magnitude estimates made during product selection. More accurate values should be used when planning manufacture.

and D is close to 10^{-5} cm^2/sec, so

$$k = \left(\frac{D}{\alpha}\right)^{2/3}\left(\frac{h}{\rho\hat{C}_p}\right) = 0.01\left(\frac{h}{\rho\hat{C}_p}\right). \qquad (4.2\text{–}10)$$

Because k is 10^{-3} cm/sec in liquids, $(h/\rho\hat{C}_p)$ must be around 0.1 cm/sec.

The trouble comes because ρ and $\hat{C}_p$ differ for different chemical species. Thus $(h/\rho\hat{C}_p)$ may be similar for many liquids, but the heat transfer coefficient h will have values which vary more widely, as suggested by the estimates in Table 4.2–1. Still, our advice at this early stage is to use these estimates, and to postpone until later the more accurate but more complex dimensionless correlations for these quantities.

EXAMPLE 4.2–I   A DEVICE THAT ALLOWS WINES TO BREATHE

David Anderson, an alumnus of the University of Minnesota who says he got poor grades in chemistry, is the buyer for the best wine store in Minneapolis-St. Paul. In an interview, he described the practice of allowing wines – especially red wines – to "breathe" by exposing them to air before drinking. The following is a summary of his ideas on this subject.

Wines need to breathe to reduce the "hard" and "soft" tannins that naturally occur. These are reduced by exposing the wine to oxygen. Reactions of oxygen with polyphenols may also be involved. Such exposure enhances flavor, allowing both fruit taste and aroma to be more clearly perceived.

The amounts of oxygen required by wines vary widely. Uncorking the bottle for 15 minutes before drinking is useless, but exposing wines to excessive oxygen over time will turn them into vinegar. Mr. Anderson suggests one good way is to pour the wine from the bottle into a decanter, leaving any residue behind. Several pours between decanters is better. Mr. Anderson even has one friend who pours the wine into a larger glass, covers the glass with his hand, and gives the wine a good shake.

In thinking about this, we must remember that wine is unstable, and can be destroyed by mistreatment. Heat and excessive oxygen are major enemies. Corks often fail, so that different bottles from the same vineyard and the same vintage may be different. Because of this, Mr. Anderson applauds the use of plastic corks. Moreover, like humans, wines can become fragile with age: a fine 25-year-old Bordeaux can degrade while it sits in the glass.

Estimate the aeration needed in a product that can let wine breathe in only a few minutes.

SOLUTION

A problem like this begs for experimentation. Our experiment used four wines:

Montepulciano, *CITRA*, Italy (1996) $6/1500 ml
Cabernet Sauvignon, *Castillero de Diablo*, Chile, (1996) $8/750 ml
Gamay, *Chateau la Charge*, France (1995) $12/750 ml
Zinfandel, *Folie a Deux*, USA (1997) $18/750 ml

Aeration can make a startling difference, especially for cheap, freshly opened wine. We tried four methods:

1. Open the bottle for 15 minutes.
2. Decant into an open pitcher and let the wine sit 2 hours.
3. Decant the wine fast three times, entraining air.
4. Shake for 10 seconds in a large glass.

Then we tested the wine.

The first method is useless, as David Anderson suggested. The other three give roughly similar taste improvements. Still another method, putting wine in a blender, aerates it excessively, dramatically reducing flavor.

These results are consistent with estimates based on mass transfer. To see why, we make a mass balance on oxygen transferred into the wine:

$$\frac{\text{oxygen transferred, } M}{\text{wine volume, } V}$$
$$= \{[\text{mass transfer coefficient, } k][\text{partition coefficient, } H]$$
$$\times [\text{concentration in air, } c_1]\} \times [\text{area per volume, } a][\text{time}, t].$$

In symbolic terms,

$$m/V = \{kHc_1\} At,$$
$$\frac{m/V}{Hc_1} = kAt = NTU,$$

where NTU is the number of transfer units. NTU is a dimensionless quantity, commonly used in the design of absorption towers.

Estimates for our four experimental methods are summarized in Table 4.2–2. The mass transfer coefficients in this table were estimated in different ways. For the first two methods we assumed a value of 10^{-3} cm/sec, as suggested in Table 4.2–1. The aeration achieved by the third method was estimated by using correlations for bubbles and for the fourth from the penetration theory of mass transfer. These

TABLE 4.2–2 Four Methods for Letting Red Wine Breathe

Method	Mass Transfer Coefficient (cm/sec)	Wine Area/Volume (cm²/cm³)	Time (sec)	NTU
Uncork bottle for 15 minutes	10^{-3} [a]	3/750	900	0.004
Decant into pitcher 2 hours before serving	10^{-3} [a]	100/750	7200	1
Decant three times, entraining 0 (5 cm³) air	20×10^{-3} [b]	0.4 [b]	120	0.5
Shake 30 cm³ of wine twenty times in a 300-cm² glass	7×10^{-3} [c]	300/30	10 [c]	0.7

[a] This typical value can be estimated from Table 4.2–1 or from the free convection caused by ethanol evaporation.
[b] Estimated from observed 5 cm³ of 0.1-cm bubbles entrained in each decantation.
[c] Estimated from penetration theory with a penetration time of one shake (0.5 s).

methods were preferred because of the small bubbles and short contact times, which give unusually large mass transfer.

The estimates in Table 4.2–2 are reassuringly consistent. Opening the bottle for 15 minutes gives less than 0.01 transfer units: it does not give enough oxygen transfer. The other three methods give around one transfer unit, apparently enough to aerate the wine, but not so much as to cause a lot of flavor loss. These three methods supply close to one transfer unit even though they have radically different areas, times, and mass transfer coefficients.

Thus our selected design – whatever it is – should give around one transfer unit. Any engineering solution meeting this specification will give about the same performance. However, we believe that the success of a "wine breather" will depend strongly on the aesthetics of the aerator's design. It is our own confidence in our lack of ability in this aesthetic area that inhibits us in developing this product.

EXAMPLE 4.2–2 A PERFECT COFFEE CUP

We have been asked by a chain of upmarket coffee shops to develop an improved coffee cup. The current cup has a volume of about 200 cm³ and a total surface area, including the top and bottom, of 200 cm². The improved cup should keep the coffee at an optimal "drinkable" temperature, estimated to be 51°C, for as long as possible. Data for coffee cooling in the current covered and uncovered cups are shown in Figure 4.2–1.

Generating ideas has led to three major directions for an improved cup:

1. A better insulated cup;
2. A cup with its own, self-contained heater; and
3. A cup with a thermal reservoir that melts around 50°C.

Select among these ideas to see which merits further development.

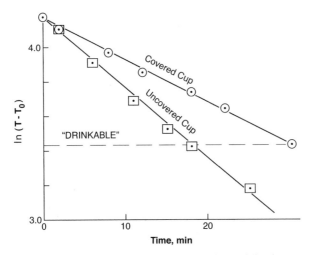

Figure 4.2–1. Cooling a Coffee Cup. The slope of the data on this semilogarithmic plot is a measure of the overall heat transfer coefficient of the cup. In these experiments, the ambient temperature was 20°C.

SOLUTION

We begin by trying to explain the data in Figure 4.2–1. From an unsteady energy balance on the coffee, we find

$$\frac{d}{dt}[\rho \hat{C}_v V T] = U A(T_0 - T),$$

where ρ is the coffee's density; $\hat{C}_v (\doteq \hat{C}_p)$ is its specific heat capacity; V is its volume; t is the time; U is the overall heat transfer coefficient averaged over the coffee; A is the coffee's surface area, including the contact with the cup and with the air; and T and T_0 are the temperatures of the coffee and the surrounding air, respectively. This equation is easily integrated to give

$$\frac{T - T_0}{T(t = 0) - T_0} = e^{-t/\tau},$$

where the characteristic time τ is

$$\tau = \frac{\rho \hat{C}_v V}{U A}.$$

This equation implies that the temperature of the cooling coffee should vary exponentially with time, which is consistent with the experiments shown in the figure.

Note that the slope of the data shown is a measure of τ and hence of the overall heat transfer coefficient. In particular, for the uncovered cup,

$$\tau = 24 \text{ min},$$

$$\frac{\rho \hat{C}_v V}{U A} = \frac{(10^3 \text{ kg/m}^3)[(1.18 \times 10^3 \text{ J})/(kg \,°K)] 2 \times 10^{-4} \text{ m}^3}{U(0.02 \text{ m}^2)(60 \text{ s/min})},$$

$$U = 57 \text{ W/m}^2 \,°K.$$

For the covered cup, τ is 40 min and U is 17 W/m^2 $^\circ$K. In this example, we have some experimental data available, and so estimates for heat transfer coefficients are not required.

We are now in a position to select among the three ideas suggested above. We first consider the idea of a better insulated cup. This idea aims at decreasing U, and hence reducing the slope of the data in Figure 4.2–1. Note that the coffee does not stay at the drinkable temperature, but rather passes that temperature more slowly. Still, there are many routes to this improved insulation, some of which are cheap. This simple, powerful idea merits further development.

The second idea, a cup with its own heater, is weaker. In this idea, we could let the coffee cool uncovered until 50°C, and then let the heater take over. From our data, we see that the heater must provide power of about

$$Q = UA(T - T_0),$$
$$= \frac{57 \text{ W}}{\text{m}^2 \, ^\circ\text{K}}(0.04 \text{ m}^2)(51\text{--}20^\circ\text{C}),$$
$$= 70 \text{ W}.$$

An average c-cell battery provides about 1 W, so we will need a lot of batteries for this concept. It just does not make sense.

The third idea is to build a coffee cup with a thermal reservoir. The physical form of this reservoir is not clear at this stage; it might replace the insulation with a layer of a substance that melts near 50°C. A coffee cup with such a thermal reservoir would behave very differently to normal cups. When the hot coffee is added, the reservoir substance would melt, cooling the coffee. When the coffee has cooled to the compound's melting temperature, the compound would begin to freeze. The heat of fusion would keep the coffee near the melting temperature until all the compound was frozen. The coffee would then cool in the normal way.

Although this could potentially work, we need more information about possible substances to make an informed judgment. One group of substances are higher hydrocarbon waxes, such as pentacosane ($C_{25}H_{52}$) which melts at 53°C with a heat of fusion ($\Delta \hat{H}_f$) of 220 kJ/kg, or beeswax, which melts at about the same temperature with a heat fusion around 180 kJ/kg. We can find the mass m of the substance required to keep the coffee at the melting temperature for a time t from the relation

$$m\Delta \hat{H}_{\text{fusion}} = Qt = UA(T - T_0)t.$$

For example, for beeswax, we can keep the coffee at 53°C in a covered cup for 20 minutes if

$$m\frac{180 \times 10^3 \text{ J}}{\text{kg}} = \frac{17 \text{ J}}{\text{sec m}^2 \, ^\circ\text{K}}(0.02 \text{ m}^2)(53\text{--}20^\circ\text{C})(1200 \text{ sec}),$$
$$m = 0.07 \text{ kg}.$$

We will need a few ounces of beeswax. Even though there are still major unknowns in this product, the idea merits further development. One important consideration will be how to ensure that heat is quickly transferred from the freezing wax to the hot coffee.

4.3   Less Objective Criteria

This chapter is about exploring ways of choosing between our better looking frogs in the hope that one of them will one day grow into the prince of our dreams. In Sections 4.1 and 4.2 we saw how the traditional tools of chemistry and engineering can be used in this selection process. We made use of chemistry and engineering to compare like with like. In this effort, it is vital not to underestimate the usefulness of chemistry and engineering training in making commercially significant decisions. This is probably the most important difference between chemical product design and parallel methods of inventing new consumer products, in which the technical input may be less central.

However, in finally making our selection, subjective decision making will also be necessary. This may take two forms. First, we are likely to need to "compare apples with oranges," that is, to make decisions between objective, but different, criteria. Often this will involve balancing cost against performance. For example, we may wish to use a new form of exhaust catalyst which will improve air quality but which will also be expensive. Chemistry and engineering will allow us to estimate how well the catalyst will perform and how much it will cost, but cannot help us in balancing these two criteria.

Second, we are likely to have to evaluate genuinely subjective issues – what do people like, how much do they care, and so on. For example, in the high-level waste disposal example of Section 3.4, the public reaction to the proposed solution is a subjective criterion. The technically attractive idea of making the waste into concrete – "grout" – was penalized because of a likely negative public response, although it performed well on all the objective criteria.

Introducing these additional elements into product selection is the subject of this section. The methodology we suggest is that of the concept selection matrix, already introduced in Section 3.4 for concept screening. The basic idea is the same: generate a handful of important criteria on which to judge the ideas; weight the criteria according to their perceived significance; and score each idea for each criterion, often relative to a benchmark that may be an existing product or well-established technology. An overall score for each item is then produced by summing the products of weighting factors and scores over all criteria. The product selected is that with the best score.

The difference between the selection matrix used here and that used in the previous chapter comes in the level of detail demanded. In Section 3.4, we were interested in reducing perhaps around twenty frogs to four or five promising candidates to be considered in greater detail. The selection phase is where that greater detail is required: all the remaining ideas are promising, so distinguishing between them becomes harder. Each number in the selection matrix (the weighting factors and the scores) must be determined with as much accuracy as possible. Where appropriate, detailed calculations should be carried out; market research may be desirable; canvassing for opinion beyond the core team is certain to be necessary. Equally, a more thorough estimate of inaccuracies is required: how much do scores vary within the core team? To what extent do external experts disagree? Once we are down to the handful of ideas left at the selection stage, we should apply a degree of effort and rigor that would be wasted on the many frogs of Section 3.4.

As chemists and engineers, we will be far less comfortable with this subjective decision making than with the quantitative analysis made possible by, for example, thermodynamics or chemical kinetics. However, we need to choose between apples and oranges and we need to worry about how people respond. This consideration of subjective criteria is necessarily woolly. The best we can hope to achieve are some useful guidelines to help us apply the selection matrix. In order to do this, we first consider important aspects of selection matrices to supplement what we learned in Chapter 3.

WHEN TO MAKE SUBJECTIVE JUDGMENTS

Because we dislike the idea of vague subjective choices, we will postpone them as much as we can. We would like to go as far down the road of objective evaluation as possible before venturing into the quagmire of subjective choice. There is no point in worrying about whether the public will like a product if it is bound to be uneconomic anyway.

The earliest and the most important point at which subjectivity becomes unavoidable is in determining the criteria to be used in the selection matrix. Inevitably cost and technical feasibility will be important, but what else matters? Do your customers mind noise, humidity, cold? Are they concerned with environmental issues? Ultimately you will need to decide, but it is useful to remember two points. First, do not leave anything out – canvas opinion widely and be aware that what is dear to your heart may be of no concern to those who will buy the product. Second, bear in mind that there may be more than one answer. For example, the British probably will not be overly worried by humidity in their homes, but this might be a crucial concern in Kuwait.

The next point at which subjectivity cannot be postponed is comparing unlike objective criteria – apples versus oranges. This element of the decision making process will be manifested primarily in the weighting factors assigned to the various objective criteria. Again, remember that the answer depends on the audience and that your own instincts may not fit with the views of your market. A Frenchman might tell you that the most important thing about a red wine is that it is well balanced, an Australian that it should be "big," and a student that it should be cheap. None of these answers is more right than another. To supply the correct product for your target market, you must assign a weight to each criterion that will depend on your customer.

Finally, you will need to include the truly nebulous – what is prettier, pleasanter, smoother, more environmentally friendly? As chemists and engineers, we naturally tend to play down the importance of these issues because we have no satisfactory means of dealing with them. We can delay their consideration until last for this reason, but we make a great error to ignore the purely subjective or to underestimate its importance. The French car, Citroën 2CV, sells because of its "look" as well as because of its price and reliability. Such considerations are less important in the case of chemicals, which are often somewhat divorced from the consumer, but nonetheless cannot be ignored.

HOW TO MAKE SUBJECTIVE JUDGMENTS

We cannot completely answer this question; all we do here is outline a few basic points worthy of attention. Let us first consider how to choose the criteria for our selection matrix. As we have already mentioned, this is probably the most crucial step in subjective decision making and the easiest to overlook. Leaving out something of critical concern to the customer is likely to be disastrous. Three aspects are key:

1. *Use independent criteria.* For example, crampons for mountain climbing need to be light and strong. However, these are not independent criteria: more strength implies more weight. A better single criterion would be that the crampons should be made from a material with a high strength/weight ratio.
2. *Avoid repetition.* This will imbalance the scores by invalidating the weighting factors you have carefully constructed. To a very large extent this boils down to ensuring that each criterion is tightly defined. We can use the high-level waste as an example again. Safety and public response will both be important. However, the public perception of each project will also depend on how safe it is. If we do not take this into account, the weighting we are giving safety will be artificially raised.
3. *Most important, use a complete list of criteria.* The list must include all the most important factors – and remember it should be what is important to the customer, not to you. This can be difficult to achieve. Canvas opinion as widely as possible, before finalizing the criteria to be used. Information from market research and independent experts is particularly valuable at this point.

Having produced a list of criteria, we must weight each and then score the competing ideas. Inevitably this is a somewhat arbitrary process, so absolute significance is not assigned to the results.

At the point of selection it is usually worth putting more effort into canvassing opinion, doing research to firm up the scores, and performing sensitivity tests. Only a few, reasonably attractive ideas remain and the next stage in product design becomes substantially more expensive. Backtracking is easy and often valuable up to the point of selection – mistakes then become costly.

WHY WE USE SELECTION MATRICES

There are clearly many other possible ways of making the decision we require. We could get someone else to choose – a manager or a customer. We could simply go with our intuition (Remember Caxton!). We could prototype and test each of our ideas. This will be reliable but costly and, crucially, slow – time to market is often the prime determinant of a product's success. Neurofen dominates its market, though it is more expensive than ibuprofen, which is chemically equivalent but arrived later in the shops.

Although selection matrices are far from perfect, we do believe that there are several advantages in using the systematic approach they imply.

1. The selection stage is the point of no return. Costs escalate rapidly once the best one or two ideas are chosen, as an extensive program of testing, prototyping, and market research becomes necessary. Simultaneously, this will be the point of fierce management review, so we must be able to justify our choice to those outside the core team. The decision matrix approach forces you to stop and consider seriously each stage of your decision making process – and this at a point when there will be considerable pressure from managers keen to minimize time to market. It also ensures that you thoroughly document your deliberations.

2. The need to weight and score each idea forces the core team both to efficiently pool its resources and to search for external input. The need for justification of criteria, sensitivity tests, and so on will make the consulting of experts and customers inevitable. The use of numerical scoring tends to make it harder for a single personality to dominate the core team's deliberations.

3. The selection stage is the last point at which we can sensibly combine aspects of different ideas to produce an improved product. The separate scoring of the different criteria ensures that the strengths and weaknesses of each idea will be very obvious and opportunities for enhancements by combination should stand out. Good, but imperfect, princes may be combined into improved models.

Despite the systematic approach offered by the concept matrix, there is no need to lose out on intuition. It is always possible to override the conclusions of the matrix – but it is better to realize you are doing it. In reality, many business decisions are made on gut feeling and products are launched more on a hunch than on sound reasoning. For example, throughout the nineteenth century, railway investments almost always lost money. However, the investors continued to be attracted by the romance of travel and progress – entrepreneurs and engineers

TABLE 4.3–1   Selecting a Prince of Wales

Parameter	Weighting Factor	Prince Charles	Hapsburg Prince	Romanov Prince
Health	0.5	5.5	10	1
Looks	0.5	5.5	1	10
Total		5.5	5.5	5.5

Note: The incumbant and two alternatives have similar scores but for very different reasons.

continued to make large fortunes by building railways. Using a concept matrix gives one the opportunity to exercise informed intuition.

EXAMPLE 4.3–1   MONARCHY SUBSTITUTION

As we have described the selection of princes, we take as an example the replacement of the British Prince of Wales. First we must establish our selection criteria. There are really only two things required of a prince. First, good health: monarchy is a symbol of stability and this is badly undermined if royalty keep dying off. Second, good looks: a prince must meet the demands of TV appearances, damsels in distress, and so on. We will weight these criteria equally. Next we need some ideas. We could do worse than look at the unemployed royal families of Europe; the most noble are the Hapsburgs and the Romanovs. How do they match up to our requirements? We will use the current market leader, Prince Charles, as our benchmark, giving him the average score of 5.5 on each criterion. On looks, the Romanovs score big. The Hapsburgs sadly suffered from a notorious chin (perhaps as a result of the severely limited breeding opportunities available to European royalty). On health, the Hapsburgs perform excellently. The penultimate Emperor, Francis Joseph, died in 1916 at the age of 87, having ruled for 68 years. Sadly, the Romanovs score poorly on health. Their particular genetic defect was hemophilia. The Tsarevich Alexis was so infirm that he often had to be carried at state functions by the sailor Darevenko. We might produce the selection matrix in Table 4.3–1. Interestingly, all three options have the same scores, but for widely different reasons. Changing these scores implies changing the weighting factors, or introducing new criteria for a successful prince.

EXAMPLE 4.3–2   THE HOME VENTILATOR

In older houses, the mean residence time of air is often around 40 minutes. Demands for draft reduction and energy efficiency mean that houses are increasingly well insulated and sealed, so the mean residence time can be 12 hours in modern houses. This is good for comfort and efficiency, but it is unhealthy. The air inside well-sealed houses contains excess CO_2, CO, radon, and formaldehyde. (This last evaporates from carpets and drapery, particularly when they are new.) The formaldehyde is often the first to reach dangerous levels. However, cigarette smoke is a very efficient scavenger of radon, so in the houses of smokers radon may be an even greater problem. Currently, there is no legislation regulating the turnover time of air in homes. Anomalously, fresh air replacement is required every 3 hours for laboratory animals in the U.S. It seems likely that legislation for humans will soon catch up.

Assuming that complete air replacement is required every 3 hours, is it possible to combine the economics and comfort of a tight house with the health advantages of a leaky one?

SOLUTION

The simplest answer is to open a window! This of course destroys much of the benefit, but it is easily achieved and will have the desired health benefit. We will take as our benchmark product an automatic window opening device that monitors air flow into the house and keeps the opening of a window at a level such as to maintain the flow necessary for a 3-hour exchange of air. This device will be cheap and easy to retrofit into "tight" houses.

Can we do better than this? Another approach is to exchange the air through a heat exchanger to minimize energy loss. This will be more comfortable, reducing the drafts to be expected as a result of an open window. A disadvantage, particularly in very cold climates, is that the incoming air will be very dry whereas that leaving the house will have a significant water content. Thus the house will dry out and become uncomfortable. One possible solution, to evaporate water, will be expensive because of the latent heat required. An alternative is to devise a ventilator that exchanges both heat and water vapor from the outgoing air stream into the incoming one. Although such units have not been commercially constructed, available literature suggests that polyimide membranes could be used for such an application.

We have two alternative products to compare to our benchmark: heat exchange with water evaporation and simultaneous heat and mass exchange. We will start by using an engineering analysis to estimate the feasibility, cost, and energy saving for each product. We will then use a decision matrix in order to assess the product's value relative to the benchmark, including some subjective criteria. We perform our analysis for the cases of Minnesota and Cambridge. In what follows, plain text refers to Minnesota and italics to Cambridge.

SETTING UP THE PROBLEM. We assume an average house in Minnesota has a floor area of 250 m². *In Cambridge, houses are smaller, perhaps half the size.* Thus we find:

Volume of air in house = 250 m² × 3 m ceilings = 750 m³ *(375 m³)*.

Number of moles of air in house, n, is given by

$$\Delta H = n \underbrace{\left(\frac{7}{2}R\right)}_{\text{gas heat capacity}} \Delta T = 29{,}000 \text{ kJ } (7{,}300\,kJ),$$

where $(7/2)R$ is an estimate of the gas heat capacity.

This is our energy cost per complete air exchange in terms of specific heat. But what about humidification and latent heat?

Now we assume the external relative humidity is 0% *(60%)*. That inside is the same, a comfortable 20%, in both Minnesota and Cambridge. The energy required to evaporate water to reach this humidity is

$$\Delta H(\text{vap}) = \left(y_{\text{H}_2\text{O}}^{\text{in}} - y_{\text{H}_2\text{O}}^{\text{ex}}\right)\Delta \underline{H}_{\text{vap}} = 9000 \text{ kJ}$$

where $\Delta \underline{H}_{\text{vap}}$ is the specific enthalpy of evaporation of water; and $y_{\text{H}_2\text{O}}^{\text{in}}$ and $y_{\text{H}_2\text{O}}^{\text{ex}}$ are the mole fraction of water vapor inside and outside the house, respectively.

In Cambridge, the external mole fraction of water is

$$y_{\text{H}_2\text{O}}^{\text{ex}} \approx 0.6(0.0087) = 0.0052,$$

and the internal mole fraction is

$$y_{\text{H}_2\text{O}}^{\text{in}} = 0.2(0.023) = 0.0046,$$

where 0.0087 bar and 0.023 bar are the saturated vapor pressures of water at 5°C and 20°C, respectively.

Therefore, we expect on a "normal" day in Cambridge that there will be no need to replenish the house's water content. Of course, the weather fluctuates a lot. On some days it may be desirable, but on average the cost will be low.

Back in Minnesota, the heating costs for the tight house, with air turnover every 12 hours, are

$$\frac{(29{,}000 + 9000) \text{ kJ}}{12 \text{ hr}[3600 \text{ (sec/hr)}]} \times (24 \times 30)\frac{\text{hr}}{\text{month}} \times \frac{\hat{=}0.03}{\text{kwhr}}$$
$$= \hat{=}19 \text{ per month } (\hat{=}3.65 \text{ per month}).$$

The heating costs for a house with a window opener, with air turnover every 3 hours, are

$$\frac{(29{,}000 + 9000) \text{ kJ}}{3 \text{ hr}[3600 \text{ (sec/hr)}]} \times (24 \times 30)\frac{\text{hr}}{\text{month}} \times \frac{\hat{=}0.03}{\text{kwhr}}$$
$$= \hat{=}76 \text{ per month } (\hat{=}14.60 \text{ per month}).$$

We expect that a heat exchanger recovers 70% of the specific heat. The heating for a house with such a heat exchanger plus humidification and a 3-hour air turnover costs

$$\frac{(0.3 \times 29{,}000) + 9000}{3 \times 3600} \times (24 \times 30) \times 0.03$$
$$= \hat{=}35 \text{ per month } (\hat{=}4.40 \text{ per month}).$$

Finally, we expect that the heat and water vapor exchanger recovers 70% of both heat and water vapor. The cost of heating a house with both heat and mass exchanger and a 3-hour air turnover is

$$0.3 \times 76 = \hat{=}23 \text{ per month } (\hat{=}4.40 \text{ per month}).$$

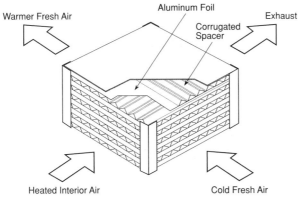

Figure 4.3–1. Schematic of a Domestic Crossflow Heat Exchanger. Warm, stale inside air is used to heat cold, fresh outside air. This heating, which reduces domestic energy use, takes place across sheets of aluminium foil.

MAKING THE HEAT EXCHANGER. We are trying to build a unit with 70% efficiency for heat recovery. We will assume a countercurrent unit with parallel plate geometry for simplicity: a schematic is shown in Figure 4.3–1. In such a device the molar flow rate of air, GA, is

$$GA = \frac{33,500 \text{ mol}}{3 \text{ hr} \times 3600 \text{ (sec/hr)}} = 3.1 \text{ mol}^{-1} \ (1.6\,mol^{-1}).$$

Total heat transfer rate across the whole exchange surface is

$$Q = GA\tilde{C}_p(T_{\text{out}} - T_{\text{in}}) = 3.1 \times (7/2R) \times 21 = 1900 \text{ W } (440\ W).$$

Now consider the heat flux per unit area, Q:

$$Q = UA\Delta T;$$

where ΔT is the temperature drop across the heat exchanger surface, a constant 9°C (5.5°C) throughout the unit. The overall heat transfer coefficient, U, is given by

$$U = \frac{1}{1/h_{\text{ex}} + 1/h_{\text{wall}} + 1/h_{\text{in}}},$$

where h_{ex}, h_{wall}, and h_{in} are the heat transfer coefficients of the exhaust air, the wall, and the entering air, respectively.

The exchange surface can be constructed of 1-mm-thick aluminium sheet: this will have a large h_{wall} and a negligible thermal resistance. Thus the h_{wall}^{-1} term can be ignored. The heat transfer coefficient for the incoming air, h_{in}, will equal that for the exhausting air, h_{ex}. Thus

$$U = \frac{1}{2}h_{\text{ex}}$$

but

$$h_{ex} = k_T/\delta,$$

where k_T is the thermal conductivity and δ is the film thickness over which heat transfer is occurring.

If we assume that the flow is laminar and that we have parallel plate geometry, then δ is approximately 1/4 of the channel dimension in the heat exchanger (i.e., halfway to the middle). Let us assume a channel dimension of 3 mm, so

$$U = \frac{1}{2}h_{ex} = \frac{1}{2}\left(\frac{k_T}{1/4 \text{ channel}}\right) = \frac{1}{2}\frac{0.026 \text{ W m}^{-1}\text{k}^{-1}}{1/4(0.003)\text{ m}} = 17 \text{ W m}^{-2}\text{K}^{-1}.$$

The area, A, required for the exchanger is now

$$A = \frac{1900}{17 \times 9} = 12 \text{ m}^2 \text{ (5 m}^2\text{)}.$$

This type of simple parallel plate heat exchanger, a well-established technology, can be built for around ≐6 per m². The production cost will be about ≐72 (≐30) and we might expect to sell the units for perhaps eight times this, a few hundred pounds.

Before we proceed, we should check that the flow is indeed laminar. To do so, we recognize that the volume of heat exchanger is $12 \times 0.003 = 0.036$ m³.

If the exchanger is cubic, the cross sectional area available for flow in each direction is about

$$= 0.5(0.036)^{2/3},$$
$$= 0.055 \text{ m}^2.$$

The superficial velocity of air v is given by

$$v = \frac{750 \text{ m}^3}{3 \text{ hr} \times 3600 \text{ (sec/hr)}} \times \frac{1}{0.05 \text{ m}^2} = 1.27 \text{ ms}^{-1}.$$

Thus the Reynolds number is

$$\text{Re} = \frac{dv}{\nu} = \frac{1.27 \times 0.003}{1.3 \times 10^{-5}} = 300.$$

That is, the flow is probably laminar.

MAKING THE HEAT AND MASS EXCHANGER. The equipment shown in detail in Figure 4.3–2 is similar to the heat exchanger shown in Figure 4.3–1. However, the problem is now more complex because we have simultaneous heat and mass transfer. We will use the same approach, looking at two key equations:

1. The total transfer across the whole exchanger.
2. The rate of transfer per unit area.

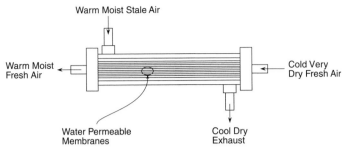

Figure 4.3–2. Schematic of a Countercurrent Heat and Mass Exchanger.
Warm, conditioned, but stale inside air heats and humidifies cold, dry,
fresh air. A water permeable membrane replaces the aluminium foil of
the previous figure.

There will now be two sets of these two key equations, one for heat transfer and
one for mass transfer.

We will assume the same geometry as for the heat exchanger, but replace the
aluminium exchange surface with a 30-μm-thick polyimide membrane. Such a
membrane is selective for water transport and has a negligible resistance to water
exchange. It is also sufficiently thin that its thermal resistance can be ignored
(as was done for the aluminium in the normal heat exchanger).

We will first look at heat transfer and examine in detail a point on the exchange
surface. The heat flux, q, at such a point is given by a combination of an enthalpy
term (heat carried by water moving across the membrane) and a conductive term:

$$q = \underbrace{C_p n_A (T - T_0)}_{\text{enthalpy term}} - \underbrace{k_T \frac{\partial T}{\partial z}}_{\text{conductive term}} .$$

We cannot separate heat and mass transport. However, we can link them by
considering the total flows across the whole exchange unit. As before, the total
heat flow, Q, is

$$Q = qA = GA\tilde{C}_p(T_{\text{out}} - T_{\text{in}}) = 1900 \text{ W}.$$

The total molar flow of water, $n_A A$, is

$$n_A A = GA(y_{\text{out}} - y_{\text{in}}) = GAy_{\text{out}}.$$

Thus

$$\frac{n_A}{q} = \frac{y_{\text{out}}}{\tilde{C}_p(T_{\text{out}} - T_{\text{in}})}.$$

Notice that this is only true because we have a countercurrent exchanger with a
flat plate geometry and equal flow rates in and out. Thus both mass and heat fluxes
are the same at any point on the exchange surface.

The heat flux now becomes

$$q = \left(\frac{y_{\text{out}}}{T_{\text{out}} - T_{\text{in}}} \right)(T - T_0)q - k_T \frac{\partial T}{\partial z},$$

$$q \left[1 - \frac{y_{\text{out}}(T - T_0)}{(T_{\text{out}} - T_{\text{in}})} \right] = -k_T \frac{\partial T}{\partial z},$$

$$q[z]_0^\delta = \frac{(T_{\text{out}} - T_{\text{in}})}{y_{\text{out}}} k_T \left\{ ln \left[1 - \frac{y_{\text{out}}(T - T_0)}{(T_{\text{out}} - T_{\text{in}})} \right] \right\}_{T_m = T_b + 1/2\Delta T}^{T_b},$$

$$q = \frac{k_T}{\delta} \frac{(T_{\text{out}} - T_{\text{in}})}{y_{\text{out}}} ln \left(1 + \frac{y_{\text{out}} 1/2\Delta T}{T_{\text{out}} - T_{\text{in}}} \right), \quad \text{choosing } T_0 = T_m.$$

Because $y_{\text{out}} < 0.005$ and ΔT is of the same magnitude as $(T_{\text{out}} - T_{\text{in}})$, we have

$$ln \left(1 + \frac{y_{\text{out}} 1/2\Delta T}{T_{\text{out}} - T_{\text{in}}} \right) \approx \frac{y_{\text{out}} \Delta T/2}{T_{\text{out}} - T_{\text{in}}},$$

and so

$$q \approx \frac{k_T}{\delta} \frac{\Delta T}{2} = U\Delta T.$$

That is, the result is identical to the previous (no mass transfer) case. This means that the enthalpy term in the heat transport equation is small compared with the conductive term.

Once again,

$$A = 12 \text{ m}^2$$

for 70% efficiency of heat exchange.

Now consider the mass transfer. We want to calculate y_{out} for an exchange area of 12 m². In other words, we want to check that the efficiency of water exchange is at least 70%.

The molar flux in a binary system is given by the following mass transfer equation:

$$n_A = (v c_A + j_A) = (n_A + n_B)y_A - cD(\partial y_A/\partial z),$$

where water is A and air is B. But y_A is close to zero because water is dilute in air. In general both c and D are temperature dependent. Thus in principle we need to incorporate the spatial variation of (cD). However, the range of temperature in the heat exchanger is small ($\pm 15°C$), so (cD) does not vary much. Thus we integrate to find

$$n_A = \frac{p}{RT} D \frac{\Delta y}{2\delta}.$$

We now return to the equation for total molar flow across the exchanger:

$$n_A A - GAy_{\text{out}} = GA(0.005 - \Delta y),$$

$$\frac{p}{RT} D \frac{\Delta y}{2\delta} A = GA(0.005 - \Delta y).$$

TABLE 4.3–2   Summary of Objective Factors for House Ventilation

Parameter	Costs	
	Minnesota	Cambridge
House with 3-hour air turnover	$\hat{=}76$ per month	$\hat{=}14.60$ per month
+Heat exchange (and humification)	$\hat{=}35$ per month	$\hat{=}4.40$ per month
+Heat and mass exchange	$\hat{=}23$ per month	$\hat{=}4.40$ per month
Unit price for heat exchange unit	$\hat{=}500 +$ installation	$\hat{=}250 +$ installation
Unit price for heat + mass exchange unit	$\hat{=}500 +$ installation	$\hat{=}250 +$ installation
Payback time: heat exchange unit	12 months	25 months
Payback time: heat + mass exchange unit	9 months	25 months

Note: Payback times are months of Winter.

For an N_2-H_2O mixture at 308 K, $D = 0.256$ cm^2 s^{-1}. Thus

$$\frac{101325}{8.314 \times 278} \times 0.256 \times 10^{-4} \frac{\Delta y}{2 \times (1/4 \times 0.003)} \times 12 = 3.1(0.005 - \Delta y),$$

$$\Delta y = 0.0013.$$

The efficiency of water exchange in our 12 m^2 polyimide exchanger is

$$\frac{0.005 - 0.0013}{0.005} \times 100\% \approx 70\%.$$

Thus both heat and mass exchange efficiency targets are met simultaneously for this 12 m^2 polyimide exchanger.

We summarize the results of these calculations in Table 4.3–2. This summary recognizes that the cost of polyimide membranes is likely to be greater than the price of aluminium sheet. However, most of the cost of exchanger units can be expected to be in the construction, so we can assume that the cost of production for the heat and mass exchanger is similar to that for the heat only exchanger already estimated. On the purely objective criterion of economic benefit, both heat and heat/mass exchangers look like winners in Minnesota. The situation is less clear in Cambridge: the heat exchanger may be viable, but the heat/mass exchanger has little advantage.

However, this analysis is not sufficient to make a decision. There are other, subjective factors to consider. For simplicity, we restrict ourselves to three additional ones: health (as air quality), comfort, and noise. We now have five criteria: heating cost, capital cost, health, comfort, and noise. We have separated heating cost and capital cost rather than include a single economic criterion because we might wish to give slightly higher weight to the reduction of heating bills on environmental grounds. We will now use these five criteria to illustrate how a decision matrix can help in the final selection process.

Table 4.3–3 exemplifies the decision matrix we can draw up. The values in this table include both the engineering results of Table 4.3–2 and the subjective factors. These merit discussion.

TABLE 4.3–3   Decision Matrix for House Ventilation

Parameter	Weight	3-hour Exchange (Benchmark)	Heat & Mass Transfer	Heat Transfer & Humidification
Heating Cost	0.3	5	10 (7)	8 (7)
Capital Cost	0.2	5	3 (3)	3 (3)
Health Benefit	0.2	5	5 (5)	5 (5)
Comfort	0.2	5	10 (7)	8 (6)
Noise	0.1	5	2 (2)	2 (2)
Total		5.0	6.8 (5.3)	5.8 (5.1)

"Health" scores identically for all products: it is redundant. We have included it for completeness and it would certainly be needed if a larger range of products was being considered. It probably should be dropped.

The results for comfort are more interesting. Both exchangers give humidity control when the outside is dry. However, the heat and mass exchanger also gives the potential for humidity reduction (in summer). Comfort is not such a large problem in Cambridge, where conditions are less extreme and so the scores are more moderate. However, both exchangers require a pumping system, which may result in irritating noise. Notice that we have scored the benchmark, which is silent, 5: this inevitably leads to scale compression. We could avoid this, perhaps by scoring the benchmark differently. The conclusions from this decision matrix are pretty much what we already know. In Cambridge, the viability of these exchangers is marginal. In Minnesota, their viability looks good. The heat and mass transfer unit looks somewhat better than heat transfer with humidification.

This conclusion includes consideration of factors beyond the economic and engineering criteria. We have been forced to think hard about what matters for a home ventilation system and how much to weight each criteria. In doing so we almost certainly will have had to talk to our potential customers, component manufacturers, and so on. The next stage of development will be to build a prototype – this will be expensive. It is important to do some consultation and hard thinking first, and the discipline of the decision matrix ensures that we do these things.

4.4   Risk in Product Selection

In our focus on product selection, we must normally make comparisons between very different product options. In some cases, the products try to accomplish the same objective in slightly different ways. In these cases, our decisions are straightforward. For example, we may be making ingredient substitutions, as described in Section 4.1. We may be simply comparing performance, as in the wine aeration example in Section 4.2.

In other cases, our product selection will be complicated by a combination of objective and subjective factors. In these cases we must balance considerations such as heat transfer with more subjective questions such as health and comfort. As explained in Section 4.4, we must then make estimates of the relative importance of these factors, and decide on ratings for each product alternative. This process, which is similar to that used for idea screening in Chapter 3, often makes technically trained persons uncomfortable. It makes us as authors uncomfortable. Still, those making these estimates, including us, are reassured that the results so obtained often seem reasonable. The basis for our product selection is exposed.

In some cases, however, we may not be sure that all aspects of all product options will work. For example, we may be uncertain of the details of a chemical synthesis, or unclear as to how the synthesis can be accomplished at greater than laboratory scale. We may not know if the mixing can be as quick as we need, or if the heat transfer to a fluid of unusual rheology will be as fast as predicted from standard correlations.

In these cases, we are selecting between products with varying degrees of risk. We consider this risk in our selection in two ways. First, we must judge how serious a particular risk is, and how much this risk will affect our product. Often, this risk assessment means that we are making estimates of extra product development money and of additional product development time. Second, we will want to reduce the risk as much as we can, perhaps by some quick experiments. Such risk reduction seeks to manage our chances of success. These two considerations, risk assessment and risk management, are the subject of the following paragraphs.

RISK ASSESSMENT

The assessment of risk most commonly involves three steps. First, we must identify and catalog all risks. Second, we must decide if these risks can be estimated with engineering tools, or if they simply generate uncertainty. Third, we must compare our possible products in terms of both cost and time.

The identification of risk begins by making a list of any possible difficulties. Making this list includes the same techniques used in the generation of ideas. We must discuss the risks with our core team and with others in our organization, especially those in manufacturing, who up to now may have been less involved with product design than other groups. We must again contact our customers, especially the lead users of our product. We may again check with consultants, looking for problems outside our own organization's experience.

Armed with this list of risks, we on the core team must then choose a probability and a consequence of each risk. The probabilities should range from zero to one. If the probability of the risk happening is negligible, it should have a score of less than 0.3. If the probability of the risk is significant, it should have a score around 0.5. If the probability is very likely, it should be scored above 0.9.

Those in nuclear engineering have much more exact definitions for assigning these probabilities. Although we do not think these are required for most chemical products, we do think that these definitions are worth mentioning. For example, a probability less than 0.01 is defined as happening once in 10,000 years.

A probability of 0.7 will happen once in 10 years, that is, within the normal lifetime of the chemical process equipment.

In addition to the risk's probability, we must assign consequences of the risks. Again, we choose a scale which varies from zero to one. Defining this scale is important: a low score means that the consequence is minor, and a high score means the consequence is major. Thus if the consequences of a risk are small, it should have a score of less than 0.3. If the consequence is significant, it should have a score around 0.5. If the consequence is severe enough to kill the project, it should have a score greater than 0.9.

The evaluation of the probability and the consequences should largely be the responsibility of the core team. As before, we suggest that each individual member assign values separately, and that the team reach its joint assessments by consensus. Assigning values as individuals forces every member of the core team to think about each risk, and to consider what he or she knows and what he or she does not. That prepares each member for the next core team meeting. In the meeting itself, core team members can reassess their scores, often by seeking information from those with different expertise. Again, we stress our belief that evaluation by consensus is better than evaluation by simple averaging. Consensus demands discussion.

Once we on the core team have agreed on the probabilities and the consequences, we must define the level of each risk:

$$[\text{risk level}] = [\text{risk probability}][\text{risk consequence}]. \tag{4.4--1}$$

We will normally decide to keep only those risks that are above a specified level, perhaps 0.5. The risk level of such "significant" risks is clearly arbitrary, and again must be decided by consensus. It will also depend on the consequences of that risk.

This combination of risk and probability concerns us deeply. If we are making a comparison of similar chemical products which have relatively little risk, we find this "risk level" concept reasonable. However, if we are comparing secure technologies with those which may fail completely, then the simple combination suggested above seems questionable. We draw no conclusions from this concern; we only urge caution when one product idea really has a large chance of failure, or the consequences of failure are very high (for example, a serious explosion).

The risk levels defined above do focus our attention on the key concerns with each possible product. We must now separate these into categories, those that can be clarified by chemical or engineering analysis, and those that cannot be. Examples of the risk levels that can be analyzed are that a heat transfer correlation might be inappropriate, or that a chemical reaction could be slower than expected, or that the product could be highly viscoelastic. In these cases, we can judge how serious the risk level is, using the chemical and engineering analyses such as those in the first three sections of this chapter. Often, these risks will be evaluated not by the core team, but by those specialists who are supporting the core team.

The high-level risks that cannot be clarified by chemical or engineering analysis must be evaluated by the core team itself. These risks will include the consequences of the marketplace and of politics. For example, we will be at risk if our chemical raw materials are only available from one supplier. We will be at risk when our manufacturing requires new licenses, or when we expect litigation from the local community to delay our facilities' expansion.

For these risks, we on the core team must guess the extra time and money that these risks imply. Doing so requires our judgment, based on our collective experience. For example, if we suspect heat transfer correlations are in error, we may judge that it will take one engineer six months to develop corrected correlations. If we expect a public hearing about our request for a building permit, we may anticipate a three-month delay, from experience of earlier struggles.

RISK MANAGEMENT

Having identified and tried to quantify both the risks and their potential consequences, we must decide on the appropriate response. We have two choices:

1.   Reduce the risk before proceeding with product development.
2.   Accept the risk and proceed.

The first strategy is the more traditional method of risk management in product development; it is outlined below. It will often require developing several different ideas in order to establish which is the most effective. This method works well, but it is slow and can be expensive.

In fact, speed and especially time to market are important considerations in chemical product development. Because of the emphasis on speed, many companies are moving toward a project-based organization. Risk reduction involves research, experiments, process design, and market testing, all of which are time consuming. Delay is in itself a commercial risk; a competitor may reach the market first and even if their product is inferior, they are then likely to gain most of the market share.

There is an analogy here with mountain climbing on glaciers. Climbing courses will, quite correctly, put a strong emphasis on safety; the extensive use of ropes, ice screws, and snow stakes will be encouraged. However using these safety measure takes time, and the safety gear is heavy, slowing progress further. The longer one spends on a glacier, the more chance there is of being caught in an avalanche; this is increasingly true later in the day as the sun softens the snow. To a considerable extent safety is speed and delay can be fatal. Similarly, delay in product design, even for good reasons of risk management, can kill off our product idea. For this reason, having evaluated the risks inherent in our different product ideas, we may wish to proceed directly to the manufacturing stage. The risk evaluation may affect our solution, but we may decide against further research at this stage, for reasons of speed.

This approach often makes good commercial sense. We may risk losing less money by proceeding with a product idea that might not work, or at least may take up larger development resources than expected, than by delaying and risking loss of market share. If it is a simple question of balancing financial considerations,

---

TABLE 4.4–1   Risk During New Drug Development

Status	Chemical Status	Quality Status	Risk
Preclinical efforts	Major process work needed	Few methods available	High
Phase I clinical trials	Laboratory procedures available	Analytical development necessary	High
Phase II to Phase III clinical trials	Pilot plant production	Analytical methods in place	Moderate
Late-stage clinical trials	Production process fixed	Methods validated	Low
Mature product	Plant process available	Quality control key	Low
Generic drug	Patents available	Methods sometimes available	Moderate

Source: Charles M. Boland, Cedarburg Laboratories, quoted in *Chemical and Engineering News*, Feb. 14, 2000.

---

we are happy with this approach. We are more uneasy when we are balancing the risk of delay against considerations such as health, environment, or safety. This is perhaps why these areas are heavily regulated to ensure companies are not tempted to cut corners in order to speed product development.

The final aspect of risk concerns our efforts to resolve the most major pitfalls before we become overcommitted. This is sometimes phrased as two guidelines:

1. If the risk is high, keep the investment low. As the risk decreases, raise the investment.
2. Break the risk into increments, deciding where you will stop work if unsuccessful.

For example, if you are developing a new device that carries out a chemical change, you could proceed in three steps. First, repeat your engineering estimates, removing the most major simplifications, and using physical property estimates that are pessimistic. If the idea still looks good, build a realistic model, at a scale convenient for the laboratory. Use experiments with the model to see if your estimates are reasonable. If things look good even now, then build a pilot scale model with which you can get the data for a final selection. We will illustrate this and other ideas of risk management in the following example.

As a second example, consider how risk wanes during development of a new drug, as suggested schematically in Table 4.4–1. Risk is highest during the initial development, when the efficacy of the drug is uncertain. If clinical trials continue to be positive, risk drops as the drug's synthesis is better established and the drug's value becomes more certain. Risk is smallest for a mature drug, produced at carefully monitored quality. Ironically, risk may increase for a generic drug, because a new manufacturer will know the patents, but not the trade secrets, for the drug's manufacture.

EXAMPLE 4.4–1 POWER FOR ISOLATED HOMES

In many European countries, electricity companies are required to provide power to homes at a fixed connection fee and standard cost per unit consumed, regardless of the homes' remoteness. Laying many kilometers of cabling to connect a single house to the national grid is clearly uneconomic.

Investigate alternative sources of electric power for isolated homes.

SOLUTION

We briefly review how one might follow the design template suggested in this book to reach the stage at which risk should be considered.

NEEDS. We will not attempt to provide electric heating, but will aim to fulfil all other normal domestic requirements, such as cooking, lighting, cooling, and so on. A little research indicates typical power requirements to average 3 kW, with a peak loading of 15 kW (mainly a result of cooking). This provides our specification.

IDEAS. There are a very large number of ways of generating electricity, some obvious (such as hydroelectric power), others more bizarre (natural gas from manure). Idea generation and initial screening might lead one to consider four leading contenders: a diesel generator, wind power, solar power, and a fuel cell.

SELECTION. For us, as the electricity provider, the primary selection criterion is going to be cost, both in terms of capital and the running cost of providing the specified power. (Remember we can only charge the standard, national rate.) Clearly our solution must also be acceptable to the consumer.

DIESEL GENERATOR. A 15-kW generator costs around $6500. Running cost can easily be estimated from the price of gasoline if we assume an efficiency of 30%. The combustion of gasoline is

$$1/8C_8H_8 + 25/16O_2 \rightarrow CO_2 + 9/8H_2O \quad \Delta H = -733.8 \text{ kJ mol}^{-1}.$$

At the efficiency given, this suggests that

$$\frac{733.8 \text{ kJ/mol}^{-1}}{1/8(104 \text{ g/mol})} \times 800 \text{ g L}^{-1} \times 4 \text{ L/gal} \times (0.3)/\$1.60 \text{ per gallon}$$
$$\approx 34{,}000 \text{ kJ per } \$.$$

In a year, we need

$$3 \text{ kW}(3600 \text{ s/hr}) \, 24 \text{ hr/day} \, (365 \text{ days/year}) = 95 \times 10^7 \text{ kJ/year}.$$

Thus our fuel cost for the diesel generator is

$$\frac{95 \times 10^7}{34{,}000} = \$2800/\text{year}.$$

Although noise might be an issue, we can expect a diesel generator to work well. Because it is very well-established technology, we will use it as our benchmark.

WIND POWER. A 3-kW generator has 3-m-diameter blades and costs around $5000. We must make use of a battery to provide the peak power load. (This will probably be necessary anyway to smooth the uneven wind energy.) We need two such generators. Once installation and battery costs are included, we are unlikely to escape with capital costs under $20,000. Running costs will be negligible. Aesthetics could be an important issue.

SOLAR POWER. Even the most efficient solar cells only manage to convert around 10% of the absorbed solar energy. Furthermore, we can only use the cells on average 12 hours per day, less in winter when power requirements will be high. Incoming solar energy is about 100 W m^{-2}. To supply our 3-kW average power, we will need:

$$\frac{(24\,\text{hr}/12\,\text{hr})3\,\text{kW}}{[0.1(\text{kW/m}^2)]0.1\,\text{efficiency}} = 600\,\text{m}^2 \text{ of solar panels.}$$

Again, we will require a battery, the efficiency of which we have ignored. In winter, the power available will be lower. Solar panels cost around $100/m^2, giving a capital cost of at least $60,000. We reject this idea on economic grounds.

FUEL CELL. Although the running cost will parallel that of the diesel generator, we can now avoid the limitations of Carnot efficiency and hope to reach 70% efficiency. Therefore running cost is

$$\$2400 \times 0.3/0.7 = \$1000/\text{year.}$$

Fuel cells are still substantially more expensive than the equivalent conventional generator: we can expect to pay at least $15,000.

A decision matrix based on the considerations just outlined might leave three contenders: the diesel generator, wind power, and the fuel cell. Before immediately proceeding with the idea that seems most attractive at this stage, we must consider risk.

In this case, the major risks are:

1. Customer acceptability, including noise and environmental considerations;
2. Regulatory acceptability, including pollution and local permission;
3. Maturity of technology; and
4. Reliability.

The third of these takes account of the effort likely to be required to bring the technology to the level where it can be slotted into our application. This risk applies primarily to the fuel cell, a relatively unestablished technology. Fuel cells operate reliably on a large scale with hydrogen fuel, but considerable uncertainty remains about running them reliably at small scale in remote locations with gasoline.

TABLE 4.4–2 Risk Assessment for Wind Power

Risk	Probability	Consequence	Risk Level
Customer acceptability	0.5	0.5	0.25
Regulatory acceptability	0.5	0.7	0.35
Maturity of technology	0.1	0.3	0.03
Reliability	0.7	0.3	0.35

Note: The most serious risks are regulatory and reliability.

The final risk listed reflects the likely costs of repairs and maintenance. Here the wind generator, vulnerable to storms, is likely to be the biggest problem.

Our risk assessment for wind power is given in Table 4.4–2. We are unsure about reducing these risks. Planning restrictions are severe in most of Europe. We cannot change the chances of wind turbines being rejected on aesthetic grounds. Similarly it is hard to alter negative consumer reactions to these unsightly and mesmeric objects. Research may help in improving reliability, but on the whole the wind power option looks risky. We are unlikely to pursue this option.

Our risk assessment for fuel cells is in Table 4.4–3. Maturity of the technology and reliability are significant risks. Both of these could be mitigated by further research. We should proceed by installing reliable diesel generators in the short term. At the same time, we may decide to pursue research into the prospects of the fuel cell option.

EXAMPLE 4.4–2 TAKING WATER OUT OF MILK AT THE FARM

Remote dairies in New Zealand can face major expenses in shipping their milk to a central processing facility, where the milk is largely made into cheese. These dairies would benefit from a method of removing only water to concentrate the milk on the farm. For a typical farm, this means reducing 4000 kg/day raw milk to about 1000 kg/day of concentrate.

Our efforts to resolve this problem have focused on four unit operations: evaporation, absorption, spray drying, and reverse osmosis. Evaporation is the best established, and is used for products such as evaporated milk and condensed milk. It requires careful energy integration. Absorption of water in inorganic and organic gels has significant problems. The inorganic gels that are selective require a lot of energy for the regeneration required for reuse.

The organic gels – such as polyisopropylacrylimide – are easily regenerated but are not sufficiently selective. Spray drying works well only with a feed of 50% solids, much more than that in raw milk. Reverse osmosis membranes foul too easily.

Thus our best idea is evaporation. From an extensive energy analysis, not included here, we decide to run the evaporator at 60°C, using 64°C steam. The steam is produced by sending the 60°C evaporated water through an electrically driven heat pump. (We should remember that a heat pump is approximately a Carnot

TABLE 4.4–3   Risk Assessment for the Fuel Cell

Risk	Probability	Consequence	Risk Level
Customer acceptability	0.3	0.5	0.15
Regulatory acceptability	0.1	0.3	0.03
Maturity of technology	0.5	0.7	0.35
Reliability	0.5	0.5	0.25

Note: These risks assume hydrogen can be handled safely.

engine run backward, using work to move heat up a temperature gradient.) The use of a heat pump reflects the fact that hydroelectric power generation is common in New Zealand, and so electricity is relatively cheap.

Within our choice of evaporation, we have three possible forms of evaporators. The first is the conventional, falling film unit, whose performance is well established and is the sensible benchmark. The second is the centrifugal evaporator, which uses centrifugal force to stabilize thin milk films and hence improve evaporation efficiency. This method works well but the equipment is expensive. The third is a membrane evaporator, where the milk films are stabilized between membranes, which can impede evaporation. This membrane method has not been carefully explored and so has considerable risk.

Select which of these ideas is best. In this selection, compare the evaporators' performance and their risk. Suggest a strategy for product development.

SOLUTION

The solution to this problem implies a total of five steps. The first step is to determine the general specifications that any evaporator must meet. The next three steps are to find the size and cost for each of the three evaporators. The final step is to consider the risk, which in this case is largely associated with the membrane evaporator.

GENERAL SPECIFICATIONS. We must first specify the general heat transfer characteristics of any successful evaporator. Doing so depends on choosing values for the physical properties of milk. Because the evaporation increases the concentrations of milk solids and nonvolatiles, the viscosity increases from 0.9 cp to around 10 cp during evaporation. We will include this change in our calculations, but will assume that other properties of the milk remain close to those of pure water. Thus the milk's density is taken as 1000 kg/m^3 and its thermal conductivity is about 0.60 W/m°K.

The total heat transferred, Q, is proportional to the mass evaporated N_1:

$$Q = UA\Delta T = \Delta \hat{H}_{vap} N_1,$$

where U is the overall heat transfer coefficient; A is the evaporator area; ΔT is the temperature difference, in this case 4°C; and $\Delta \hat{H}_{vap}$ is the specific heat of vaporization at $60°C$, here about 2430 kJ/kg. Because N_1 is 3000 kg/day or 0.035 kg/sec,

$$U A = 21 \text{ kW/°K}.$$

But

$$\frac{1}{U} = 1/h_{\text{steam}} + 1/h_{\text{wall}} + 1/h_{\text{milk}},$$

where h_{steam} is the individual heat transfer coefficient of the condensing steam, around 5000 W/mK; h_{wall} is that of the evaporator surface, typically 20,000 W/mK; and h_{milk} is that of the milk itself. We assume that this is given by

$$h_{\text{milk}} = k_T/\delta,$$

where k_T is the thermal conductivity of the milk, and δ is the milk film thickness. Thus if we can estimate δ, we know h_{milk} and hence U, and so can find the area of a particular evaporator. This will be the key parameter in our selection.

FALLING FILM EVAPORATOR. The first unit we consider is the conventional falling film evaporator. In this unit, the film of milk must spread smoothly over the evaporator surface in order to efficiently use all of the surface. Such a smooth film means that the Weber number, We, must be greater than a critical value of 2:

$$We = (\rho v^2 \delta/\sigma) \geq 2,$$

where ρ is the milk's density, v is its velocity, and σ is its surface tension. For a falling film,

$$v = \rho g \delta^2/3\mu,$$

where g is the acceleration due to gravity and μ is the viscosity. Combining

$$\delta = \left(\frac{18\mu^2\sigma}{\rho^2 g^2}\right)^{1/5}$$

$$= \left[\frac{18(0.1 \text{ g/cm sec})^2 \, 30 \text{ g/sec}^2}{(1 \text{ g/cm}^3)^2(980 \text{ cm/sec}^2)^2}\right]^{1/5}$$

$$= 0.14 \text{ cm}.$$

To make sure we have a stable film, we assume we want about twice this value, or

$$\delta = 0.3 \text{ cm}.$$

From the above, we then find that h_{milk} equals [(0.60 W/mK)/0.003 m], U is about 200 W/mK, and the evaporator area A is

$$A = 100 \text{ m}^2.$$

This evaporator area, the benchmark for our selection, is large because the temperature difference is small ($4°C$).

CENTRIFUGAL EVAPORATOR. The centrifugal evaporator uses centrifugal force to keep the milk film smooth, thin, and stable. As the milk film moves outward on the centrifuge disks, its higher viscosity caused by evaporation is more than balanced by the increased centrifugal force. Although the details of the fluid mechanics are beyond the scope of this book, the result is that the average film thickness is about

$$\delta = 25 \; \mu m.$$

Parallel with our earlier arguments, we now find that h_{milk} equals $[(0.60 \; W/m \, °C)/25 \times 10^{-6} \; m]$, U is about 5000 $W/m^2 \, °C$, and the evaporator area A is

$$A = 5 \; m^2.$$

Using a centrifugal evaporator cuts the surface area required for evaporation by over thirty times.

However, this dramatically reduced area is dearly purchased. The only serious estimate that we could obtain for building a centrifuge like this was over $50,000. This seems too expensive for most farmers. As a result, we turn to the third method for evaporation.

MEMBRANE EVAPORATION. Like the centrifugal evaporator, the membrane evaporator can sustain very thin, stable milk films during evaporation. As shown in Figure 4.4–1, the thin films are now sustained not by centrifugal force but between two thin membranes. One of the membranes is a metal foil, which transfers heat from the 64°C steam to the 60°C milk. This membrane has a heat transfer coefficient around 20,000 $W/m^2 \, °C$.

The other membrane, which separates the 60°C milk from the 60°C steam produced by the evaporation, is the barrier for the evaporation. Interestingly, its heat transfer resistance and mass transfer resistance are predicted to be neglible under

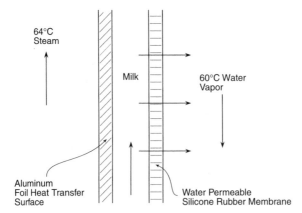

Figure 4.4–1. Membrane Evaporator. Water in warm milk evaporates across the thin membrane shown. Because the membrane is selective, volatile flavors are not lost.

TABLE 4.4–4   Risk Assessment for the Membrane Evaporator

Risk	Probability	Consequence	Risk Level	Mitigation
1. Difficult to make heat transfer membrane	0.1	0.5	0.05	Use parallel heat exchange technology
2. Difficult to make evaporation membrane	0.3	0.5	0.15	Existing data suggest, at most, required membrane area doubles
3. Cannot easily manifold the module	0.5	0.2	0.10	Can mitigate with larger steam channel
4. Evaporation flow is slow	0.5	0.2	0.10	Use larger membrane spacer in steam channel
5. Cannot sterilize effectively	0.3	0.9	0.27	Chemical cleaning preferred, but requires no dead spots

these conditions. If this is true, then the significant resistance to heat transfer must be the film of milk itself. In many membrane devices like this, the two membranes are held apart by a spacer, which fixes the thickness of the milk film. Typically, the thickness δ of this spacer is

$$\delta = 600 \ \mu m.$$

By the same arguments as before, h_{milk} is $[(0.60 \ W/m^2 \ ^\circ C)/6 \times 10^{-4} \ m]$, U is about $900 \ W/m^2 \ ^\circ C$, and the evaporator area A is

$$A = 23 \ m^2.$$

This is one fourth the area of the falling film evaporator, but eight times the area of the centrifugal evaporator. Significantly, membrane experts agree that membrane modules like this can be built for about $10/m^2$, independent of the chemical structure of the membrane used. Thus we should be able to build a membrane evaporator for less than $1000. Such a system is attractive commercially.

RISK ASSESSMENT. The three evaporators discussed above show a vivid contrast of advantages and disadvantages. The traditional thin film evaporator has the largest area because it operates with the thickest milk film. The centrifugal evaporator has a very small area but a very high price. The membrane evaporator has a moderate area and a very low price, but it may not work. Using the membrane evaporator is risky.

Five of the major risks of this evaporator are shown in Table 4.4–4. The first, that we have trouble making the heat transfer membrane, is unlikely because there are already foil-based heat exchangers on the market. The obvious strategy is to use the manufacturing procedures of these foil exchangers as a guide. The second risk, that the membrane across which evaporation occurs offers a major mass transfer resistance, is more serious. Although such trouble would be inconsistent with earlier studies of membranes with high permeability, we suspect that the membranes used in those studies may be difficult to make in the large, flat sheets needed here. However, even if the water permeability is only 20% of that reported earlier, the membrane area required increases only slightly.

The other risks depend on the design of the evaporation module. The third risk concerns the design of the inlets and outlets, and should not be especially difficult to resolve. The fourth risk reflects the concern that the evaporated water will not easily flow out of the module. This is easily mitigated by using a larger membrane spacer in the steam channel. Sterilization of the milk channel is the most severe risk. Although the membranes may not be able to stand high temperature, most farms use chemical cleaning anyway. We must ensure that sterilization is complete, without any dead spots. Even this risk, scored as the most serious, does not seem crippling. We should build a prototype and show by experiments if this new but risky idea merits selection.

4.5   Conclusions and the Third Gate

This chapter aims at selecting the best one or two product ideas to manufacture. Before this point, we have identified product needs and ideas that might fill these needs. We have qualitatively screened these ideas until we are left with five or so leading possible products. We want to choose the best one or two of these product.

How we do so has been described in this chapter. The methods used for product selection often use quick estimates of thermodynamics and rate processes. These estimates are especially useful when we are trying to improve existing products, either with new ingredients or with more responsive systems.

The challenge comes when we compare new, very different products with improved but familiar ones. In this case, we return to concept selection matrices, which attempt to weigh the relative importance of different attributes. We must also consider the importance of risk, for a new proposed product may not work as well as we hope. We must consider what new engineering and chemistry are needed to mitigate the risk.

After these efforts, we will have chosen our best potential product, and be ready to consider how it can be manufactured. But first, we must face a third management review.

This third management review, or "gate," will be by far the hardest and most critical that we encounter. Remember that the first gate was very early in the process, when we had first developed our specifications. At this point, management will not be especially critical, often because they had a role in identifying the original need. The second gate came after idea generation and sorting, when we had produced five or fewer good ideas. Again management will remain relatively uncritical, partly because they will see merit in a few of the ideas, and partly because they may not understand the chemical details of the new product ideas.

Remember that management consultants feel that these two first gates are often too casual, and that many product design projects are allowed to continue for too long.

The third gate will be the hardest because we on the core team will be asking for a lot of money. Management may not understand chemistry, but they do understand money. As before, we should write a report and prepare an oral presentation. Now, however, we are going to be critically examined, even grilled. This is the stage at which the product development is most likely to be cancelled. But if our product still looks good, we will be ready to consider how it can be manufactured. This is the subject of the next chapter.

FURTHER READING

Cussler, E. L. (1997). *Diffusion, Mass Transfer in Fluid Systems*. Cambridge University Press, Cambridge, ISBN 0521450780.

Hildebrand, J. H., Prausnitz, J. M., and Scott, R. L. (1970). *Regular and Related Solutions*. Van Nostrand Reinhold, New York, ISBN 79122670.

Louvar, J. F. and Louvar, B. D. (1997). *Health and Environmental Risk Analysis*. Prentice-Hall, New York, ISBN 0131277391.

McMillan, J. (1996). *Games, Strategies, and Managers*. Oxford University Press, Oxford, ISBN 0195074033.

Murray, R. L. (1993). *Nuclear Energy: An Introduction to the Concepts, Systems and Applications of Nuclear Processes*, 4th ed. Butterworth-Henemann, Oxford, ISBN 0750628952.

Stevens, S. S. (1985). *Psychophysics: Introduction to its Perceptual, Neural and Social Prospects*. Transaction Books, New Brunswick, ISBN 0887386431.

5

Product Manufacture

By this point, we are close to a decision on what product we will make and sell. We have identified a customer need, and we have quantified the need in terms of product specifications. We have sought a large number of ideas that could meet this need, and we have organized and edited these ideas until we have a manageable number. We have selected the best one or two ideas. Now we are close to deciding what we will manufacture.

This chapter explores three aspects leading to product manufacture. The first, discussed in Section 5.1, concerns intellectual property. Often, our new product will include some aspects of invention. In these cases, we will want to consider whether or not to seek patent protection. Patents can give us an exclusive licence to market our invention, and hence command higher prices which let us more quickly recover our development cost. In return for this exclusive licence, however, we must make a full disclosure of what our product is, and how it works. Sometimes we will decide to seek patent protection, but sometimes we will choose to keep trade secrets. This choice is the focus of the first section.

In Section 5.2, we turn to finding missing information required to realize our product. Sometimes, this information will be necessary to make sure our selected product will function as we expect. In other cases, it may be part of what is needed for any patent applications. Usually, the information must be obtained from actual chemical and physical experiments, which are almost always tedious. As a result, we are concerned in this section with planning experiments that are as efficient as possible.

The next issue, covered in Section 5.3, is to develop final specifications for the one or two possible products we are still considering. We should stress that the thinking we need in this development is adaptive, not innovative. In the early stages of product development, we encouraged innovative thinking because the resulting innovations can redefine the market. Now, we have much more invested in the potential products we have chosen. We want adaptive evolution, not revolution.

Finally, we turn to the manufacture itself. In doing so, we are building on a background in chemistry and chemical engineering. This background will include thermodynamics, chemical reaction kinetics, transport phenomena, and unit operations, which we do not review here. However, there are some aspects of this background that are not central to the manufacture of commodity chemicals, but are important for more specialized products. In Section 5.4, we discuss aspects that apply to microstructured products, such as paint and ice cream. In Section 5.5, we review aspects that apply to chemical devices, such as artificial kidneys. We defer until Chapter 6 the details of specialty chemical manufacture.

5.1   Intellectual Property

Because we are close to a decision on whether to make a specific product, we need to expand our focus from technology, and think of some broader issues of product design and development. Some of this broader perspective still centers on members of our core product team, especially those concerned with marketing. We want to reassure ourselves that the markets we identified at the start of our development process are still there, that customer needs have not shifted. We want to check again that our selected product does fit the markets which we believe exist.

This expanded thinking should also involve questions of intellectual property. Intellectual property means patents, which can prevent our competitors from making the same product. Patents can dramatically enhance the advantages of our being first to market.

The case of Jerome Lemelson serves to indicate how important intellectual property rights can be in product commercialization. At his death in 1997, Lemelson had the distinction of having the third largest number of U.S. patents held by an individual (Edison and Land are one and two). He patented over 500 inventions, but commercialized none of them. Instead he made his money by licensing his inventions or often by suing companies that he felt had infringed his rights. For example, Lemelson claimed invention of the bar code long before it became a supermarket norm. He wanted payment for each bar code used on any product. Because there is a strong incentive for companies already marketing successful products to settle this type of action out of court, Lemelson's heirs have so far not found it necessary to reach the stage of a court verdict. At his death, Lemelson was believed to have an annual income of several hundred million dollars from his patent portfolio. A charitable foundation now exists with the sole purpose of defending and pursuing the Lemelson patents and disposing of the resulting income. Patents can make a lot of money; they can also make life very difficult if someone else holds them, even if the holder has no intention of going into competition with you.

Experience in intellectual property will usually not be extensive in the core team. Thus at least some of the core team are going to talk to lawyers. Lawyers have a bad reputation with most technically trained persons. Before we discuss what we need from lawyers, we probably should discuss why they have this reputation.

We suspect that there are three main reasons why chemists and engineers dislike lawyers: money, laws, and truth. Lawyers can be better paid than chemists and engineers, partly because lawyers have been able to influence the rules by which they are paid. This is a silly reason to dislike lawyers: if chemists and engineers could influence their compensation, they would do so. "Professional registration" and "licensing" are engineering efforts in that direction. This first reason to dislike lawyers is just envy.

Second, lawyers operate within the confines of laws, the rules by which society operates. They are not necessarily concerned about whether the rules are fair; they are interested in what the rules are. Engineers yell at lawyers just as spectators yell at football referees. Moreover, when there are conflicts in rules, lawyers litigate; that is, they fight. Litigators like to fight, and some will be happy to fight – at your expense – on any side of any issue. That is what lawyers are, society's professional fighters. One aphorism attributed to a litigation lawyer says, "When the only tool you have is a hammer, the whole world looks like a nail."

The third issue is one of truth. Good lawyers tell their clients the truth, saying exactly what they mean. Engineers tend not to like that for two reasons. First, lawyers can use words that are abstruse or that have slightly different meanings from those accepted in popular speech. As authors – one English and one American – we have struggled with similar differences in meanings while writing this book. Just as we have resolved these international nuances, so we can learn to understand lawyers.

In addition, engineers and chemists will often be angered when lawyers both tell the truth and show empathy. For example, imagine an engineer in a tantrum because the lawyer asserts a patent cannot claim an early date for invention because record keeping was sloppy. After the tantrum, the lawyer may say "I can see why you are upset." The engineer may infer that the lawyer may be sympathetic and changing his mind. He is not. He is just acknowledging that incompetence in keeping records has compromised the discovery date for the patent. Most "reasons" for disliking lawyers are not rational. We can now consider the important information that they can give us.

PATENTS AND TRADE SECRETS

The intellectual property generated in product design is conveniently split into patents and trade secrets. A patent is a contract between the inventor and the government. If the inventor convinces the government, represented by the patent office, that the invention is new, then the government gives the inventor exclusive rights to the invention for a considerable time. In return, the inventor gives the public a full disclosure of what the invention is and how it works. In the language of the patent, the inventor "teaches" the public how to make and use the invention.

Patents are valuable because they grant a period when the inventor can earn higher profits and hence more easily recover development costs. The inventor can often expect to get about two thirds of the sales if the product is the first of its type to reach the market. However, the price earned during that original period can be considerably higher when the product is protected by a patent.

A patent is legal property. Just like a house or a car, patents can be owned, bought, and sold. They can be licensed, most often for fees of around 3–6% of gross sales. Moreover, a patent can be international. Although most realize that the United States, Europe, and Japan have patent systems, many do not realize that nearly 200 separate countries, from Afghanistan to Zimbabwe, also have such systems.

In contrast to patents, trade secrets are nonpublic information used in manufacturing our product. They may be a special catalyst, or important steps in activating the catalyst. They may be the organization of a microelectronic device, such as a smart chip or a debit card. They may not contain any new information, but just particular information like the key to a code. The so-called PIN number on a money card is a good personal example.

Trade secrets are not legal property, so products that depend on them are always vulnerable. This vulnerability can have two forms. Most obviously, the trade secrets can be lost with an employee who changes jobs. The employee may have developed these secrets or may simply have learned them on the job. When the employee leaves for a better job, his or her value to the new employer may strongly depend on his or her experience with the previous employer, including knowledge of trade secrets. We may wish our former employee a successful career, but we will not appreciate his or her success if it comes largely at our expense.

The second vulnerability of trade secrets can occur when one of our competitors independently discovers the secret and patents it. The ex-secret is now the competitor's property. To keep using this secret, we will need to negotiate a licence and start paying patent royalties. This can be true even if we have been using the secret for decades before the competitor discovered it.

Thus, our first decision will be whether or not to patent our product. The benefits are obvious: the patent gives us legal protection. The debits are more subtle. First, although patents do give legal rights, they can be difficult and expensive to defend. While estimates are difficult to make, many patent lawyers acknowledge that only a small fraction of infringement cases are prosecuted. Most infringement cases are settled out of court, usually on confidential terms. Thus the odds of defending a patent do not look very good, although the existence of the patent may itself inhibit promiscuous infringement.

A second possible debit of patents may occur if they are very broadly sought. For example, a drug company working on antidepressant medications could patent huge numbers of compounds before the efficacy of the compounds can really be proved. Most of these patents have little value, because most of the compounds patented will not have much medicinal benefit. However, because each patent includes a full disclosure of the chemistry involved, the entire set of patents will provide a remarkably accurate template of the company's chemical expertise and strategy for discovery that would be available to all its competitors. Anecdotal evidence suggests that this strategy for learning about competitors has been effectively used by some successful drug companies.

In making the decision as to whether to seek a patent or keep a trade secret, some companies have tried a third way. They do not patent a particular process trick, and they do not keep it as a trade secret, either. Instead, they present it as a poster at some minor technical meeting that they expect will be poorly attended.

Meetings that do not publish abstracts of posters are especially good. The company then keeps a careful, notarized record of what was in the poster, including the trade secrets. The chances that the secret will be noticed by a competitor are remote. Thus the secret essentially remains secret. If a competitor does in the future discover and patent it, the original discoverer can then refer back to the poster section. Such an earlier disclosure invalidates the patents and avoids license fees.

In general, however, we will find that the legal protection afforded by patents on carefully chosen products is well worth the effort. We will normally want to seek these patents. How we do so is the subject of the next paragraphs.

WHAT CAN BE PATENTED

Basically, the United States recognizes three kinds of patents: utility patents, design patents, and living plant patents. Utility patents are the most important for chemical products and will be discussed in detail below. Design patents, involving ornamental features of an article of manufacture, are mentioned briefly. Patents on living plants, which cover new varieties that have been asexually reproduced by the inventors, will not be discussed here.

Utility patents are by far the most common. They are granted for any useful, new and nonobvious composition of matter, article of manufacture, or process. Contrary to what many engineers say, process patents are not easier to obtain than other types of utility patents, but they are often harder to enforce. Most utility patents are complex documents. Because of their complexity, they are expensive to prepare, but they can provide broad protection.

At present, utility patents have a term of 20 years from the filing date of the earliest U.S. patent application from which the patent claims priority. This means that we will normally want the application to proceed expeditiously, especially if we hope to gain additional income from licensing the patent. Such speed is a contrast with behavior before 8 June, 1995. Before that date, the priority was based on the date of application, but the protection was extended 17 years beyond the date of issue. This meant that smart patent attorneys took as long as possible in the application process, thus effectively extending the length of protection.

Design patents are different from utility patents. Design patents are granted for 14 years for any new, original, and nonobvious ornamental design for an article of manufacture. Such a patent involves the physical, nonfunctional appearance of an invention. Such patents will not apply to a chemical compound, though they might apply to a device for chemical change, such as an artificial kidney. Because they are simple and inexpensive, patent attorneys assert that design patents are underused in product design and development.

REQUIREMENTS FOR PATENTS

We now restrict our discussion to utility patents. We are concerned with two major issues: what is patentable and how we document our invention. These two issues are summarized in the following paragraphs.

To obtain a patent we require a product that is useful, that is novel, and that is nonobvious. The "usefulness" requirement is normally easy to satisfy. It only requires any level of utility beyond casual experiments. We are not going to be interested in the commercial development of any chemical product that does not have substantial utility.

The "novelty" requirement is harder to prove, because it depends on both common sense and legal distinction. Sensibly, the product cannot have been known or used by others in the United States before your invention. The product cannot have already been patented, either by the inventor or someone else. The product cannot have been described by someone else in a printed publication.

Then things get more complicated. If the inventor has disclosed the invention in a printed publication, he has 1 year from the date of disclosure to apply for a patent within the United States. However, outside the United States and particularly in the important European market, the inventor who publicly discloses his or her invention immediately forfeits any patent rights. In the words of one patent attorney, "instant devastation is the brutal fact of life almost everywhere outside the United States."

This forfeiture of patent rights is the just cause of management paranoia about giving papers at meetings. Any public disclosure can put product development in jeopardy. A confidential disclosure agreement is only one factor of many that a court will consider in deciding if disclosure was "public." In addition, many attorneys argue against waiting after an invention to make the patent application. Thus any product developer who knows that the new product is useful and novel should move promptly to make the patent application.

In addition to being useful and novel, a patentable product should be "nonobvious." This requirement means that the differences between the new product and earlier products – "the prior art" – must be sufficient that they are not obvious to one having "ordinary skill in the art to which the invention pertains." This "nonobvious" requirement will clearly introduce substantial uncertainty into the assessment about whether the new product merits a patent. The United States Supreme Court tried to sharpen this "nonobvious" requirement by urging inquiry into three areas:

1.   The scope and content of the prior art;
2.   The differences between the art and the claimed invention; and
3.   The level of ordinary skill in the field of the invention.

The Court also listed several secondary considerations. Clearly, we must supply information to patent attorneys, but we normally must depend on their judgment.

We can aid the process of patent application by keeping careful records of our development of the product. The standard is a handwritten, bound laboratory notebook, kept in ink and witnessed weekly by a supervisor or a knowledgeable peer, who is not an inventor. This witness is asserting that the notebook has been "read and understood." However, from our own experience with patent litigation, we believe that such careful records are the exception, rather than the rule. All too many times, the records include incomplete descriptions of what was done and why, as well as scraps of paper that appear to have been added later. Such poor records are incompetent.

Moreover, the explosion of electronic files and computer-generated printouts means that most actual data are no longer carefully crafted columns of numbers written in blue ink. We recognize that computer files are more efficient and do not for a moment suggest that they be abandoned. However, for the foreseeable future, we still think that it makes sense to collect printouts on a weekly basis, to glue them into a laboratory notebook, and to have them read and witnessed by a knowledgeable peer. With these precautions, establishing the basis for a patent should be much easier.

Finally, we should stress that U.S. patent laws divide the inventive process into two steps: conception, and reduction to practice. Conception is the formation by the inventor of a definite idea of the complete invention, including every feature sought to be patented. Conception is complete when one of ordinary skill in the art could practice the invention without extensive research or experimentation. Posing a problem is not conception.

Because conception is mental, the courts also require reduction to practice, which is evidence that the invention works. Reduction to practice takes two forms: actual and constructive. Actual reduction to practice requires construction of a device or preparing a composition. Then the inventor must demonstrate that the invention fulfills its intended purpose. This actual reduction to practice often includes getting missing information, using methods such as those in the next section. Constructive reduction to practice is filing a patent application. This is normally the lawyers' responsibility.

Patent infringement is different from normal criminal offences. You cannot ring up your local police department and ask them to enforce your intellectual property rights. A patent is a legal document, but the enforcement is entirely up to the owner. Until a patent is tested in court, there is no certainty about what it really means. This can be very troublesome for small companies or individuals when faced by infringement, particularly by those with much larger resources. Normally if a patent is infringed, the patent holder is entitled to the payment of a reasonable license fee. If it can be shown that the infringement was willful, damages of two to three times this amount may be awarded. A recent case in which this has occurred is an award of over $200 million to the University of Minnesota when it was judged that Glaxo-Wellcome had knowingly infringed a University patent in producing a drug.

There are two crucial points to remember about intellectual property law. First, it is complex; when you need to get involved in patents, it is imperative that you do so with the assistance of a specialized lawyer. Second, intellectual property law has little to do with justice as it is normally understood. When asked to sum up patent law as concisely as possible, one patent lawyer described it simply as an opportunity to employ lawyers. Although we would not wish to encourage quite this level of cynicism, we must to remember that patent law is a set of rules, not a means of enforcing truth, justice, or a better way of life.

EXAMPLE 5.1–1   THE INVENTION OF THE WINDSURFER

The windsurfer was patented by Hoyle Schweitzer in 1969. Schweitzer and an engineer, Jim Drake, set up a company, Windsurfing International Inc., and did much of the early development on windsurfing boards. For many years this company produced its own windsurfers and received license fees from other manufacturers. It was not until the late 1970s that windsurfing really took off, when the craze hit Europe; its greater popularity there is sometimes attributed the longer holidays common in Europe.

At this point, windsurfer manufacture became highly profitable, and large-scale manufacturers such as BIC, F2, and Mistral became heavily involved. These companies found it irksome to continue to have to pay a royalty to Windsurfing International every time they sold a board, and they began to search for ways to circumvent the patent. Eventually BIC discovered prior art, in the form of published material prior to the patent application date. In 1958 a British inventor, Paul Chilvers, was toying with something that looked a bit like a windsurfer; a picture was found in an obscure local paper. Also, in 1964, Newman Darby built a form of windsurfer (the Darby Sailboard) in Wilkes-Barre, PA. The Darby board used a universal joint to attach the mast to the board, one of the essential elements of the Schweitzer patent.

After a hotly contested legal battle, the patent was undermined by this prior art in both the U.S. and Europe. BIC no longer pays license fees (and nor does anyone else). We do not wish to suggest any ethical conclusion as to who deserves financial benefit. We would like this example to draw attention to the uncertainties and vicissitudes associated with the ownership of a patent.

5.2   Supplying Missing Information

In Chapter 4 we saw how to make the final selection of our most promising idea. At this stage we had sufficient information to convince ourselves that the idea we chose was a winner, a prince among frogs. This information may have come from the available literature, from external experts, or from back-of-the-envelope calculations. However, the information is unlikely to be complete and rigorous. Because we are about to embark on an expensive program of product development, we had better be sure exactly how well our product is going to work.

Discovering these product details requires further research and experimentation. Up to this point we tried to minimize the work at each stage; simplified calculations have always been employed, experiment kept to a minimum, and literature research used only to establish if something is possible, with little attention to how it might be achieved. This streamlines product design, allows easy comparisons between ideas, and minimizes time to market. Now, however, detailed information is indispensable: we must confirm experimentally any information used already and fill in the many gaps in our knowledge. Our prince must be clothed and educated to become a working monarch.

It is of course extremely difficult to generalize about what missing information will be necessary and how best it can be obtained. Every project will have its own specific problems and the information available will vary enormously, depending on both the level of literature interest and your company's prior activity in this area. The minimum requirement will be experimental verification of relevant reported data. At the other extreme, a full experimental program may be necessary to demonstrate the viability of a new and untested idea.

One form of missing information, commonly required for the design of chemicals, is a synthetic pathway for the active molecules. By the time we have completed the selection stage, we will certainly have identified what our active species is, and we may well have obtained it in small quantities. We are less likely to have identified a satisfactory commercial synthesis route. Before we can move on to manufacturing, we need to identify this route. Many of the techniques of chemistry will be valuable; one which is worth further discussion is a systematic way of developing a range of reaction path strategies for the active molecule.

REACTION PATH STRATEGIES

One of the first things chemists, biochemists, or pharmacists are likely to do when considering solutions to a product problem is to start to think of possible molecules required for synthesis. In Chapter 3, we discussed how combinatorial chemistry and natural product screening can assist in this process of generating ideas for active chemical species. The molecules identified are unlikely to be readily available and usually will be complex. How can we make them?

Going backward from the target molecule to simple precursors is exactly what "the disconnection approach" to organic synthesis is designed to achieve. This approach, outlined by Stuart Warren, makes successive "disconnections" to reduce the target molecule to simple, available precursors. Each disconnection involves imagining breaking the structure of the target molecule; this breakage is the inverse of a synthetic step. A disconnection should be related to a well-established synthetic method to go in the opposite direction and so will be closely connected with the functional groups available in the target molecule.

Usually several different disconnections will be possible for any target molecule and many successive disconnections are likely to be necessary before simple precursors are reached. Thus very many alternative synthetic routes can easily result from moderately complex molecules. An experienced organic chemist will usually be able to eliminate most of them immediately as completely impractical, leaving perhaps a handful of reasonable alternatives. Sometimes, none of the potential synthetic routes will look viable. In this case the suggested target molecule can be ruled out as a useful idea; it may be highly efficacious, but if it cannot be commercially made it will be of no use as our product. Alternatively, perhaps, it could be extracted from a natural product or made via fermentation.

To illustrate the type of approach required to fill in the gaps in our information, we will give three examples.

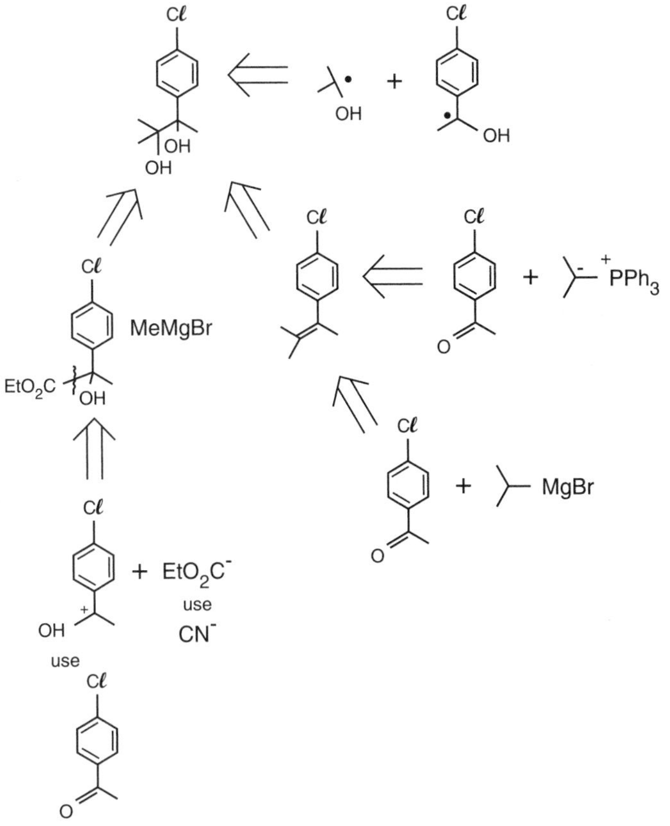

Figure 5.2–1. Phenoglycodol Synthesis. The drug, given in the upper left-hand corner of the figure, can be made by many routes. Three are sketched here. In these routes, each arrow indicates a possible "disconnection" step to simpler precursor molecules.

EXAMPLE 5.2–1   SYNTHESIS OF THE TRANQUILIZER, PHENOGLYCODOL

The structure of this species is given in the upper left-hand corner of Figure 5.2–1. Suggest several routes by which it may be synthesized.

SOLUTION

This is a fairly complex molecule and so many pathways are possible. The most obvious, also shown in Figure 5.2–1, use commercially available precursors. Which synthetic route we prefer will depend on other factors, such as cost, safety, and so on.

EXAMPLE 5.2–2   STERICALLY HINDERED AMINES FOR CO_2 REMOVAL FROM GASES

Acid gas removal from gas streams (sweetening) is a very common process in the chemical and refining industries. For example, in a hydrogen plant, methane is converted by steam reforming to hydrogen and CO_2. The CO_2 must then be removed to leave a pure product. In an existing plant, this CO_2 removal is often the bottleneck for capacity expansion. For this reason, your company would like to improve CO_2 removal from gas streams. How can you do so?

SOLUTION

Conventionally, gas sweetening is achieved by using amines, by the following reaction:

$$2RNH_2 + CO_2 \rightarrow RNH_3^+ + RNHCOO^-.$$

Reaction occurs in a gas-liquid column at low temperature (40–80°C) followed by amine regeneration at higher temperature (120°C) and low CO_2 partial pressure. This reaction requires two moles of amine per mole of CO_2 removed. In 1974, Guido Sartori, a chemist at Exxon, realized that by changing the amine, the stoichiometry of the reaction could be changed:

$$RNH_2 + CO_2 + H_2O \rightarrow RNH_3^+ + HCO_3^-.$$

Only one mole of amine is now required per mole of CO_2 absorbed. This is potentially a great improvement in efficiency. The stoichiometry can be changed by using a hindered amine, such as diisopropylamine:

The anion formed by reaction with CO_2 is destabilized because the side groups have a high electron pushing power and the bulky side chains prevent free rotation of the acid group.

However, if the amine is highly hindered, the rate of reaction becomes so slow that it is useless for CO_2 removal. What is required is a moderately hindered amine, such that a reasonable reaction rate is still achieved. This is possible with amines similar to that shown above. However, we do not know exactly which hindered amine to use. Our specifications are likely to go as follows.

1. We require a new product that will double the capacity of the old plant or reduce the size of absorption columns in a new plant. This means that we need to achieve an increase in the CO_2 carrying capacity of the absorbing liquid.
2. The rate of reaction must be at least as high as that for the conventional amines, or our capacity gain will be offset by a poor rate of absorption.
3. We want to retrofit the old plant with our new product. Therefore, operating conditions must be similar to those used currently; that is, absorption at 40–80°C, and regeneration at 120°C.
4. In an operating plant, a corrosion inhibitor, containing V^{5+}, is present in the absorbing liquid. The hindered amine must be stable in the presence of this inhibitor.

In order to develop the final product, Sartori and co-workers screened a wide range of possible hindered amines for their performance on each of these four criteria.

It is likely that they tested hundreds of possible amines to establish the optimum product. Data on around a dozen hindered amines are published in the open literature, showing the trends established as a function of the size and the chemical nature of the hindering groups.

First, Sartori established that hindered amines can indeed react with the 1:1 stoichiometry shown above. Next, he investigated the rate constants for CO_2 absorption. He found that moderately hindered amines showed almost an order of magnitude drop in rate constant relative to unhindered ones and that highly hindered amines were over an order of magnitude worse again. This led to rejection of highly hindered amines in favor of moderately hindered ones. Although the rate constant dropped by changing to hindered amines, the rate of CO_2 absorption will be a result of an expression of the form:

$$\text{rate} = k[CO_2]\,[\text{amine}].$$

Because the stoichiometry is 1:1 for hindered CO_2 absorption rather than 2:1 for conventional amines, there will be certain operating conditions under which the actual rate of CO_2 absorption is higher in the hindered amine case than in the conventional situation. Sartori and his collaborators were able to show that for some moderately hindered amines, this was the case for typical plant operating conditions. Indeed, they found that under these conditions the rate of CO_2 absorption was limited only by CO_2 diffusion into the liquid. Thus by choosing from a range of moderately hindered amines, the requirement of increased capacity without loss of rate can be met.

Having established the required degree of steric hindering for the amines, the requirements of solubility and thermal stability were satisfied by altering the chemical nature (but not the size) of the side groups. For example, the solubility in aqueous solution is usually enhanced by using alcohol side chains. For this reason, the standard unhindered amine used in conventional CO_2 absorption is diethanolamine. Undoubtedly Sartori incorporated similar chemical features into the hindered amine in the final product.

The final stage in the experimental program was to test the stability of the possible amines in the presence of the V^{5+} inhibitor. It turns out that the new product is better than conventional amines in this respect also. The new amines are now produced commercially.

EXAMPLE 5.2–3 SILVER BULLETS FOR ZEBRA MUSSELS

Zebra mussels are a freshwater mussel native to Europe, which reached North America via bilgewater in cargo ships. They have since become enormously successful, outcompeting native bivalve species and aggressively colonizing freshwater habitats. They present a particular problem to industry because of their propensity to block cooling water systems using raw water supplies, such as power station heat exchangers.

Zebra mussels feed by filtering nutritious particles from the water. It is proposed to control them by feeding them poisoned capsules, "silver bullets," the idea being to use their own filtering activity as a means of concentrating the poison. This will allow bulk toxic concentrations in the water to be many times lower than that which would be required if the poison were placed directly in the cooling water system. Hence cost and environmental damage should simultaneously be minimized. What further information would be required before this product could be commercially developed?

We must answer two questions:

1. What should the size and composition of our capsules be?
2. What concentration of capsules do we need in order to achieve a given kill rate, say 90%? How does this compare with the bulk water concentration of toxin we would need simply to poison the mussels?

SOLUTION

Question 1 is relatively simple to answer. There is a considerable amount of literature available on toxins for bivalves in general and zebra mussels in particular. One simple but promising candidate is KCl, which induces heart attacks in mussels. Another is the Ethiopian soapberry, which is rich in surfactants and is used extensively for washing clothes in Ethiopia. Waterways used for such washing are remarkably clear of molluscs. We would like to ensure that a mussel is killed if it eats a few silver bullets; we need to establish a lethal dose of each toxin. This will be done by measuring mortality after the direct administration of a given dose to the mussel's gut. We then just need to coat our lethal dose of poison in something edible, probably fat or starch. Because zebra mussels filter only fine particles efficiently, we need to fabricate our capsules in this size range. We would probably finally wish to test our silver bullets in a "live firing" exercise in which the product prototype was administered to feeding mussels and mortality checked.

Question 2 is going to be harder to answer, but crucial to the economic and environmental feasibility of the product. Establishing the fatal concentration of the poisons in solution is a relatively simple experiment. However, establishing the bulk concentration of our poisoned capsules will be harder. We need to know the filtration rate of the mussels in order to be able to calculate their probability of ingesting a silver bullet in a given water flow, such as that inside a heat exchanger. There is some literature on this, but more extensive experiments would certainly be required to mimic the conditions prevalent in industrial pipework.

It is once again clear from this example that taking a product from being a good idea, which we believe will work, to being commercially viable requires a great deal of work and, perhaps more crucially, a lot of time: mussels can take a while to die during experiments! An extensive experimental program to confirm crucial facts or fill in missing or estimated information is indispensable at this stage.

5.3   Final Specifications

At this point, we are considering making only a small number of products for manufacture. Normally, this will be one or two. The products may be either a chemical, such as a new drug for counteracting depression, or a device, such as a new catalytic convertor for reducing nitrogen oxide (NO_x) emissions.

We want to choose final specifications for this product. To begin this task, we are wise to carefully review where our work so far has led us. We should describe the chemical product we want to make. If it is a chemical, we need to specify its molecular structure, its final form, and its required purity. If the product is a device for chemical change, we must specify its physical size and shape, and its expected mode of operation. Again, we suggest that each member of the core team briefly write out these specifications, and that the core team resolve any differences by consensus.

Our final specifications should also re-examine our competition. We want to compare our new product with the best existing product. We want to identify improvements we expect, and to state how large these improvements can be. We want to restate all the assumptions that we are making, and to decide which of these assumptions involves the most uncertainty. These efforts should circumscribe our problem, and identify the technical constraints with which we must deal.

Our thinking in setting these final specifications is very different to that required earlier. Before we were especially interested in innovation, because we hoped that such "out-of-the-box" thinking would supply clues to big commercial advantages. We urged setting general specifications; we encouraged eccentric ideas; we sought surprising selections. We understood that we would reject most of these ideas. We understood that we needed to kiss a lot of frogs to find a prince.

We now want adaptive thinking, not innovative thinking. We want to improve our carefully selected ideas by slight modification, not by random invention. In many ways, we are best guided by a maxim taught as a diagnostic guide to second-year medical students:

> *When you hear hoof-beats in the street, think horses not zebras.*

For the medical students, this means that sick patients are more likely to have common diseases than exotic ones. For us, it means that our products can most likely be improved by careful, evolutionary changes, not by enormous mutations.

Setting these final specifications can often be aided by a three-step strategy. First, we define the product structure, a task that is relatively straightforward. Second, we rank the product's most important attributes, an effort that forces a review of how the product will be used. Third, we review any chemical triggers, that is, chemical stimuli which cause major changes in product properties. These three strategic steps are detailed in the following paragraphs.

PRODUCT STRUCTURE

Specifying the product structure usually involves the four chief items listed below.

1. *Chemical composition.* What is the planned product made of? If it is chemically pure, what is its chemical structure? If it is a device, how much can its composition be changed without affecting its performance?

2. *Physical geometry.* What product characteristics are fixed? Are there fixed macroscopic dimensions? Is there any unusual physics?
3. *Chemical reactions.* Does the product change chemically during use? Do acids, bases, and salts affect these changes?
4. *Product thermodynamics.* What is the product's phase? Is this phase thermodynamically stable or metastable?

We must emphasize that these concerns with product structure will apply differently to different products. Drugs will be different from shut-down battery separators. Nonetheless, we recommend this checklist, even if it only stresses the product's uniqueness.

CENTRAL PRODUCT ATTRIBUTES

We next turn to re-examining the most important attributes of the product we are almost ready to manufacture. Most often in preparation for manufacture, we will be willing to make a long list. That is not the goal here because we want to be focused. We want to choose three or fewer important attributes; we would prefer identifying only one as most important.

The long list of product attributes can often be organized under three headings:

1. *Structural attributes.* These include the product's physical properties, such as its strength and elasticity. These attributes are more important for devices than for chemical products.
2. *Equilibrium changes.* Many chemical products will show major changes in equilibrium as a consequence of altered temperature, pH, or some other process variable.
3. *Key rate processes.* The most obvious is the rate of any important chemical reaction. Less obvious but often important are rates of heat transfer, fluid flow, or diffusion, which are often manipulated by changes in interfacial area.

Again, in our search for final specifications, we should use this organization to find the most important attributes.

CHEMICAL TRIGGERS

The final strategic step seeks to identify any chemistry that makes the product become active. What frees the product from its original thermodynamic bondage? This step, more important for chemical products than for devices, usually involves a variable such as one of the following:

1. *Solvents.* These dissolve or disperse the product so it becomes useful.
2. *Temperature changes.* The most common example is regenerating a product – such as an adsorbent – by heating or cooling.
3. *Chemical reactions.* The most common occur because of pH changes or hydrolysis.

4.   *Other physical changes.* These may include pressure, detergency, and electric field.

When these three steps are complete, we should be in a position to imagine how manufacturing can occur. Before making these next steps, we turn to an example to illustrate how this strategy can improve product design.

EXAMPLE 5.3–1   FREON-FREE FOAM

Refrigerators are normally insulated with polyurethane foam. The foam is made by injecting reactive monomers into the space between the inner and outer walls of the refrigerator. Traditionally, freon was injected along with the reagents. As the reaction proceeded, the freon evaporated, producing a foam with about 95% bubbles containing freon.

The result was a very effective insulator. The properties of this insulation were used to establish standards for home refrigerators. The outside dimensions of the refrigerator became standard, so a new refrigerator would easily fit into the space occupied by the old one. The inside dimensions also became standard, so that milk bottles fit conveniently inside. The insulation required for energy efficiency was also legally restricted, with laws based on the properties of freon-containing foam.

However, when freon is released to the environment, it destroys the layer of ozone that protects the earth from excess ultraviolet radiation. As a result, an international agreement has banned the production and use of freon. To be sure, freon in insulating foam seems less abusive than freon in single use products such as hair sprays. Nonetheless, the freon in foam will eventually leak out, perhaps long after the refrigerator has been scrapped. Sensibly, polyurethane foam blown with freon is illegal, and not available.

We need to build refrigerators with the same dimensions and the same degree of insulation as those with freon-containing foam. The degree of insulation achieved depends most dramatically on the thermal conductivity of the gas in the foam's bubbles. This thermal conductivity k_T is given in W/mK by

$$k_T = (0.08/\sigma^2\Omega)\sqrt{T/\tilde{M}},$$

where σ is the molecular diameter, in Å; Ω is dimensionless and of order one, a weak function of temperature; T is the absolute temperature, in °K; and $\tilde{M}$ is the molecular weight in daltons. Thus if we replace the freon with CO_2, we find from Table 5.3–1

$$\frac{k_T(CO_2)}{k_T(CCl_2F_2)} = \left(\frac{\sigma_{CCl_2F_2}}{\sigma_{CO_2}}\right)^2 \left(\frac{\tilde{M}_{CCl_2F_2}}{\tilde{M}_{CO_2}}\right)^{1/2},$$
$$= \left(\frac{5.3}{3.9}\right)^2 \left(\frac{121}{44}\right)^{1/2} = 3.$$

The foam blown with carbon dioxide will provide only one third the insulation of the same thickness of foam blown with freon. A foam blown with nitrogen is even worse, with only one fourth the insulation.

TABLE 5.3–1   Properties of Gases Used in Insulating Foam

Gas[a]	Molecular Weight (d)	Molecular Diameter (Å)	Boiling Point (°C)
Nitrogen (N_2)	28	3.8	−196
Carbon dioxide (CO_2)	44	3.9	−79[b]
Freon 12 (CCl_2F_2)	121	5.3	−30

[a]Freon's large diameter and high molecular weight give it the lowest thermal conductivity.
[b]Sublimes.

We need a better foam. A careful search for ideas has produced many interesting alternatives, including materials made of many layers of aluminium foil. After careful analysis, however, we decide that our best choice is polyurethane foam modified in some way to reduce its thermal conductivity.

Use the strategy given above to suggest final product specifications.

SOLUTION

The three-step strategy given above suggests defining the product's structure, specifying its chief attributes and identifying any chemical triggers that make the product active. In this case, the chemical trigger is not critical, but the other steps are important.

PRODUCT STRUCTURE. Defining the structure is easy. We want a polyurethane foam containing 95% gas bubbles. The bubbles should be small to avoid free convection: free convection in any larger bubbles will compromise insulation. The idea that the bubbles could be much smaller than in the present foam is interesting, but we defer discussing this until later. There are no chemical interactions in the present foam. Again, the interesting idea of such interactions is deferred until later. In general, foams are metastable, especially if the bubbles are very small, but this should not be a major problem in this case.

THE KEY ATTRIBUTE. The foam is a good internal insulator. This key attribute is directly a result of the thermal conductivity in the foam's gas-filled bubbles. As a result, we can benefit from a review of this transport property. For a monoatomic dilute gas, the thermal conductivity k_T is given by

$$k_T = \frac{1}{3}(\text{distance between collisions})\frac{\text{energy}}{\text{volume}}\left(\frac{\text{volume}}{\text{area time}}\right).$$

The volume per area per time is nothing more than the average molecular velocity, v. For a monoatomic gas, this velocity depends on temperature, T, via the kinetic energy:

$$1/2\,mv^2 = k_B T,$$

where m is the molecular mass and k_B is Boltzmann's constant. The energy per volume, the product of the molecular concentration c and molar heat capacity $\tilde{C}_v$, is given by

$$c[\tilde{C}_v] = \frac{p}{k_B T}\left[\frac{3}{2}k_B\right],$$

where p is the pressure. We only need to estimate the distance between collisions.

There are two limiting cases of this collision distance, valid for large bubbles and for small bubbles. For larger bubbles, the distance is the mean free path, λ, that a gas molecule travels before it collides with a second gas molecule. This mean free path is related to the volume per molecule:

$$(\pi/4)\sigma^2\lambda = V/n,$$
$$= k_B T/p,$$

where σ is again the molecular diameter, V is the bubble volume, and n is the number of gas molecules in the bubble. Solving for λ and combining with the above, we find that

$$k_T = (1/\sigma^2)\sqrt{T/m}.$$

This variation of the thermal conductivity with molecular size and weight is equivalent to that presumed in the problem statement. It is the variation that lets us estimate how much poorer CO_2-blown foam would be compared with freon-blown foam.

This large bubble result is dramatically different than that for small bubbles. For small bubbles, the gas velocity and the gas energy per volume are unchanged, but the distance between collisions is different. For small bubbles, this distance is proportional to the bubble diameter. Unlike in large bubbles, where gas molecules collide with each other, a molecule in a small bubble bangs from one point on the wall to another. As a result, the thermal conductivity is now

$$k_T = dp\sqrt{k_B/2mT},$$

where d is the bubble diameter. Note how different this result is from the previous equation. Whereas k_T varies with the inverse square root of molecular weight for both large and small bubbles, k_T increases with temperature in large bubbles but decreases with temperature in small bubbles. More importantly, the thermal conductivity is independent of pressure and bubble size in large bubbles, but it is proportional to the product (dp) in small bubbles. Thus we can make a better freon-free insulating foam by having small bubbles or a low gas pressure.

SETTING FINAL SPECIFICATIONS. To complete our product specifications, we must decide what is a large bubble and what is a small bubble. From the above, we see that this difference depends on the Knudsen number Kn, the ratio of the mean free path λ, and the bubble diameter d.

$$Kn = \lambda/d,$$
$$= (4/\pi)\sqrt{k_B T}/p\sigma^2 d.$$

When $Kn \ll 1$, we have intermolecular collisions, and hence large bubbles. When $Kn \gg 1$, we have molecule-wall collisions, and hence small bubbles.

We want small bubbles. Although we can try to make these mechanically, we will find it difficult to get bubbles smaller than $1\,\mu$m, not small enough to be "small." The reason is that the surface energy of such bubbles is high, so that some of the bubbles tend to grow at the expense of others. This process is sometimes called "Ostwald ripening."

However, we could make the bubbles behave as if they were small by reducing the gas pressure, and hence raising the Knudsen number. The product designers who were involved in making a better foam did just this by a very clever invention. They blew rigid polyurethane foam with carbon dioxide in the normal way, under established reaction conditions, but they blew it into a bag made of metal foil. The bag is essentially completely impermeable to all gases. Just before the bag was sealed, the designers added a spoonful of sodium hydroxide to the bag. The sodium hydroxide reacted with any available CO_2, which slowly diffused through the foam to react. It turns out that a chemical trigger is involved in our product manufacture after all.

The result was a foam, initially the same as any other CO_2-blown foam, but which got better with time. Eventually, as the gas pressure got lower and lower, the foam became a better insulator than the original freon-blown foam which it replaced. The final product specification that the new foam must have a thermal conductivity no higher than freon-blown foam is met.

EXAMPLE 5.3–2   BETTER BLOOD OXYGENATORS

For open heart surgery, we must use a machine to bypass the patient's heart and lungs while the heart is being repaired. The machine must move the blood at roughly the normal rate, which is relatively easily accomplished with a pump. It must perform the same function as the lungs, a more difficult task: it must add oxygen and remove carbon dioxide. In almost all cases, oxygen addition is more difficult to accomplish than carbon dioxide removal, so that blood oxygenators are normally designed by using oxygen transfer as the benchmark.

We can oxygenate blood by using many familiar chemical engineering operations. For example, we could oxygenate blood in a packed tower, letting blood trickle downward over Raschig rings while air flows upward, countercurrently to the blood. This type of operation is not attractive for two reasons. First, any free interface between air and blood tends to cause clots, just as any open cut on our hands tends to clot. Blood clots can cause strokes. As a result, past designs of blood oxygenators tend to carry out the oxygen transfer across a membrane. Originally, silicone rubber membranes were used that offered significant resistance to mass transfer. More modern designs use microporous hydrophobic membranes that offer no significant resistance to mass transfer.

The second reason that blood oxygenators cannot use conventional equipment such as packed towers is that the volume of blood required to start up such a packed tower would quite literally drain the patient white. The tower could be

a) Flow Across a Helically Wound
   Bundle

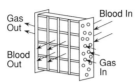

b) Flow Across a Rectangular
   Bundle

c) Flow Along a Crimped
   Flat Membrane

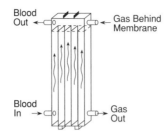

Figure 5.3–1. Current Commercial Blood Oxygenators. The designs in a) and b) use beds of air-filled microporous hollow-fiber membranes. The third design, c), uses a crimped, flat, microporous membrane.

started with blood transfusions. Unfortunately, these carry the ever-present risk of infecting the patient with HIV or hepatitis. We must use the smallest blood oxygenators that gives enough oxygen transfer.

Thus, we need a mass transfer device that offers the greatest amount of mass transfer per volume across a microporous membrane. Some designs are shown in Figure 5.3–1. Originally, the membranes used were flat. Later, to get more area, the membranes were corrugated, like furnace filters or the air filters in automobiles. Now, blood oxygenators usually use hollow-fiber membranes, which give the best performance yet achieved.

Imagine that we want to build a new hollow-fiber blood oxygenator that outperforms other models. To do so, we want to maximize the oxygen transferred per blood volume. We know that the oxygen flux per volume J is given by

$$J = Ka(c_1^* - c_1),$$

where K is the overall oxygen mass transfer coefficient, based on a liquid side resistance; a is the membrane area per volume; c_1^* is the blood oxygen concentration at saturation, kept constant by using excess air; and c_1 is the actual oxygen concentration in the blood. In this case, K is dominated by the individual mass transfer coefficient in the blood, k. Thus, our problem is simple: we must select the oxygenator design that maximizes ka.

SOLUTION

In this example, the key is the product's structure, epitomized by the product ka. There are no real chemical triggers; we just need a big ka. We begin this selection

---

TABLE 5.3–2   Mass Transfer Correlations Across Hollow-Fiber Membranes

Flow Geometry	Flow Range	Correlation
Within fibers	$Sh > 4$	$Sh = 1.62 Gr^{1/3}$
Outside and parallel to fibers	$Gr < 60$	$Sh = 1.3(\frac{d_e^2 v}{vl})^{0.9} Sc^{1/3}$
Outside and across fibers	$Re > 2$	$Sh = 0.4\,Re^{0.8}\,Sc^{0.33}$
Outside and across fiber fabric	$Re > 2$	$Sh = 0.8\,Re^{0.49}\,Sc^{0.33}$

Note: Dimensionless groups defined as follows: Graetz number, $Gr = d^2 v / Dl$; Sherwood number, $Sh = kd/D$; Reynolds number, $Re = dv/\upsilon$; and Schmidt number, $Sc = \upsilon/D$. Variables defined as follows: d is fiber diameter; v is average blood velocity; D is oxygen diffusion coefficient; l is hollow-fiber length; k is mass transfer coefficient in blood; and υ is the kinematic viscosity in blood.

---

by considering the area per volume a. For the hollow fibers, we expect

$$a = \frac{\text{fiber area}}{\text{oxygenator volume}},$$

$$= \left(\frac{\text{fiber area}}{\text{fiber volume}}\right)\left(\frac{\text{fiber volume}}{\text{oxygenator volume}}\right),$$

$$= \left[\frac{\pi d l}{(\pi/4)d^2 l}\right](\phi),$$

$$= \frac{4\phi}{d},$$

where d is the fiber diameter, l is the fiber length, and ϕ is the volume fraction of fibers in the module, normally around 0.5. Commercially available microporous hollow fibers typically have a diameter of about 300 μm. Thus, a is reasonably circumscribed, and any advantages will come from the mass transfer coefficient k.

Some of the correlations that are reported for the hollow-fiber mass transfer coefficient are given in Table 5.3–2. In these correlations, the mass transfer coefficient, k, is given as a function of many variables, in particular the fluid velocity, v. Although this velocity can vary dramatically with the geometry of the hollow fibers, the velocity per length in blood oxygenators is normally fixed:

$$v/l = 1/\text{sec}.$$

Higher velocities usually imply higher shear, which can damage the blood.

We can now look at three special geometries of hollow-fiber oxygenators. In every case, we will look at the Sherwood number, for the largest Sherwood number means the largest k, and hence the fastest oxygenation. For flow inside the fibers, no matter how the fibers are arranged, we have

$$Sh = 1.62\left(\frac{d^2 v}{Dl}\right)^{1/3},$$

$$= 1.62\left[\frac{(300 \times 10^{-4}\,\text{cm})^2\,1\,\text{sec}^{-1}}{10^{-5}\,\text{cm}^2/\text{sec}}\right]^{1/3},$$

$$= 4.5.$$

In this estimate we have chosen the diffusion coefficient D as 10^{-5} cm^2/sec, a value typical of water, not blood. Blood's higher viscosity will tend to reduce D; and the reactivity of oxygen and hemoglobin will tend to increase the effective value of D; but these changes are relatively minor.

As a second alternative, we consider flow outside and parallel to the hollow fibers. In this case, the correlations in the literature vary widely, presumably because it is difficult to manufacture fibers that are evenly spaced. Using the correlation shown, we have

$$
\begin{aligned}
Sh &= 1.3\left(\frac{d^2 v}{vl}\right)^{0.9}\left(\frac{v}{D}\right)^{1/3} \\
&= 1.3\left[\frac{(300\times 10^{-4}\,\text{cm})^2\, 1\ \text{sec}^{-1}}{10^{-2}\,\text{cm}^2/\text{sec}}\right]^{0.9}\left(\frac{10^{-2}\,\text{cm}^2/\text{sec}}{10^{-5}\,\text{cm}^2/\text{sec}}\right)^{1/3} \\
&= 1.5,
\end{aligned}
$$

where we have assumed the kinematic viscosity, v, is close to the value of water of 0.01 cm^2/sec. The value we calculate is significantly less than that for flow within the hollow fibers, so this geometry is not attractive.

Our third alternative looks at flow outside but perpendicular to the hollow fibers. From our experience with flow outside but parallel to the fibers, we expect that the fibers should be as evenly spaced as possible. One good way to do so is to weave the fibers into a hollow-fiber fabric. In this case,

$$
\begin{aligned}
Sh &= 0.8\left(\frac{dv}{v}\right)^{0.49}\left(\frac{v}{D}\right)^{0.33}, \\
&= 0.8\left(\frac{300\times 10^{-4}\,\text{cm}\ 4\,\text{cm}/\text{sec}^{-1}}{10^{-2}\,\text{cm}^2/\text{sec}}\right)^{0.49}\left(\frac{10^{-2}\,\text{cm}^2/\text{sec}}{10^{-5}\,\text{cm}^2/\text{sec}}\right)^{0.33}, \\
&= 30,
\end{aligned}
$$

where we have assumed that the hollow fibers are arranged in a bed whose depth is 4 cm, so the blood velocity is 4 cm/sec. This looks like the best geometry by far.

When we look at the selection of commercial blood oxygenators shown in Figure 5.3–1, we see that most current successful designs do use blood flow across beds of air-filled hollow fibers. We are not likely to improve on these geometrics, especially because the blood flows are carefully controlled to avoid excessive shear. To be commercially innovative, we will need to select a design that has additional advantages beyond maximizing oxygen transfer per oxygenator volume. Because such a new product may be hard to design, we should consider canceling this project.

5.4   Microstructured Products

We next turn to the manufacture of specific products: microstructured products in this section, and chemically based devices in the following section. In principle, these sections should be unnecessary, because the basic ideas important to these products are covered in core courses in chemical engineering and chemistry. However, in practice, we have found that some ideas involved are not emphasized in

TABLE 5.4–1   Average Properties of Milk

Component	Concentration (wt. %)
Water	82.0
Fat, including saturated and unsaturated fatty acids	4.0
Protein, which emulsifies the fat	3.4
Lactose	4.8
Other solids	0.8

Note: Milk is an oil-in-water emulsion, with about 3×10^9 globules/cm^3 of around 4 μm diameter. The overall composition is shown.

TABLE 5.4–2   Typical Formulations for Latex Paints

Component	Exterior (wt. %)	Interior (wt. %)
Pigment (TiO$_2$), to scatter light	20	11
Extenders, including clays	16	34
Polymers, for rheology control	18	16
Latex, to form coating	25	10
Defoamers, Dispersants, Buffers, etc.	4	3
Water	17	26

Note: Latex particles average 0.5 μm; smaller sizes give higher gloss, but larger sizes give greater opacity and better flow.
From Stokes and Evans, 1996.

these core courses, and so they merit a brief review. In this review, we will stress only the concepts that most commonly cause confusion.

We begin with microstructured products, of which ice cream and paint are good examples. Such products have properties like "creaminess" and "hiding power" that depend on the products' structure at the scale of 1–10 μm. For ice cream, a key variable is the size of the ice crystals, which are around 8 μm; for latex paint, a key variable is the diameter of the latex particles, around 2 μm. Although the molecular structure of these products can be important, similar microstructures made from different molecular species often have similar product attributes.

Microstructured products are chemically complex, as shown by the typical compositions of milk and paint given in Tables 5.4–1 and 5.4–2. These products are not the single phases common for commodity chemical products. They cannot be characterized by variables such as molecular weight or boiling point or octane number. Microstructured products like these are commonly not in an equilibrium state, but are metastable colloids.

The metastable state of many microstructured products means that their properties are not only a function of their current condition, but also of their processing. In more scientific terms, their properties are a function not only of state variables, but also of the path by which they reached that state. For example, ice cream newly

TABLE 5.4–3 Key Ideas for Microstructured Products

Intellectual Basis	Key Concepts	Products Where Important
Thermodynamics	Electrochemical potential	Water softeners
	Micelle formation	Soap and detergents
	Ostwald ripening	Ice cream
Kinetics	Colloid stability	Paint, bearnaise sauce
	Colloid inversion	Suntan lotion, butter
Rheology and Mixing	Dimensionless groups	Ice cream, floor wax

Note: Many of these ideas, which are parts of colloid chemistry, are infrequently stressed in introductory chemistry and chemical engineering courses.

purchased in the store at $0°C$ and 1 atm is very different from ice cream that is melted and then slowly refrozen to $0°C$ and 1 atm.

In spite of this major effect of process path, we can still explain many properties of microstructured products with familiar tools such as thermodynamics and kinetics. Some of these tools are listed in Table 5.4–3. In the following paragraphs, we will discuss them in more detail, and so facilitate our efforts to manufacture microstructured products.

THERMODYNAMICS

Our discussion of the thermodynamics of microstructured products depends strongly on chemical potential. Chemical potential is commonly remembered as a terrifying muddle of partial derivatives, manipulated without much physical insight to pass long-forgotten exams. Unfortunately, we can explain microstructured products most easily in terms of chemical potential.

In qualitative terms, the chemical potential is the energy of a particular species. The particular energy is the Gibbs free energy per mole, and so includes both contributions of the enthalpy and the entropy, but that need not concern us yet. The chemical potential is increased by heating the system or by putting it under higher pressure, but pressure normally has little effect for condensed phases. The chemical potential is reduced, often dramatically, by dissolving the species in a solvent. This dissolution is limited if solute and solvent are incompatible, if the solute is somehow chemically racist, preferring its own kind.

These ideas of chemical potential are quantified by equations like the following (already mentioned in considering solubility parameters):

$$\mu_1 = \mu_1^0 + RT \ln x_1 + \omega x_2^2, \tag{5.4–1}$$

where μ_1 is the chemical potential itself, in joules per mole; μ_1^0 is a reference value, a "chemical potential in the standard state" R is the gas constant; T is the temperature; ω is a measure of the heat of mixing; and x_1 and x_2 are the mole fractions of solute and solvent. (In this case the standard state is pure "1" at the T and P defined.) Note that the logarithmic term in Equation 5.4–1 is negative, and represents the reduction in chemical potential caused by dilution for an ideal mixture. For an endothermic dissolution, ω is positive, and signals the solvent's tendency toward unmixing, toward chemical racism.

A common alternative to Equation 5.4–1 is for a solvent "1" dissolving a high molecular weight polymer "2," for which

$$\mu_1 = \mu_1^0 + RT \ln \phi_1 + RT\phi_2 + \chi\phi_2^2, \qquad (5.4\text{–}2)$$

where ϕ_1 and ϕ_2 are now the volume fractions of solvent and polymer; and χ is again related to a heat of mixing. Again, the logarithmic term is negative, a reduction in the free energy of the small solvent "1." Again, the term containing χ is related to the heat of mixing. The term ($RT\phi_2$), often called the lattice correction, is the increase in free energy resulting from the polymer's large molecular size. The polymer can't mix as freely as small molecules can because its various units must stay attached, one to the other.

These basic ideas of chemical potential can explain colloidal behavior such as micelle formation and Ostwald ripening. Micelle formation is a characteristic of soaps and detergents, exemplified by sodium stearate:

$$CH_3(CH_2)_{16}COO^-Na^+$$

and sodium dodecyl sulfate:

$$CH_3(CH_2)_{11}SO_3^-Na^+.$$

Molecules such as these have a hydrophilic "head" – the ionic part – and a hydrophobic "tail" – the greasy hydrocarbon. In aqueous solution, these species agglomerate into micelles. These aggregates have the entangled hydrophobic tails in the core and the hydrophilic heads forming a skin that shields the tails from the water.

Such micelle structures can minimize the chemical potential of the soap. Moreover, the hydrocarbon core can dissolve other species that normally would be largely insoluble in water. For example, the solubility of cholesterol in water is increased 30,000,000 times if bile salts are added. Bile salts, the body's detergent, form micelles with cholesterol at the center. Such agglomeration and entrapment are often called solubilization.

Micelle formation and solubilization can be rationalized in terms of chemical potentials. Although a variety of models for this process have been suggested, the simplest, "phase separation" model is sufficient for our needs. This model assumes that micelles are a separate phase. Once micelles begin to form, any additional solute goes into this new, micellar phase. Micelle formation begins at the solute's critical micelle concentration, x_{1c}, when

$$\mu_1(\text{micelles}) = \mu_1^0 + RT \ln x_{1c}. \qquad (5.4\text{–}3)$$

The mole fraction of any solute not in micelles remains at the fixed concentration x_{1c}. This is just like a normal solubility product. Any solubilized material will go into the micelles present at an amount proportional to $(x_1 - x_{1c})$.

Micelles are a thermodynamically stable phase. As they become larger, containing more and more solubilized material, they can become thermodynamically stable microemulsions. More frequently, they become metastable, held back from fast phase separation by the same hydrophobic core and hydrophilic skin that stabilized the micelles. Such metastable emulsions, which include milk, liquid cleaners, and paints, will often remain metastable for years. We will return to this question of metastability later in this section.

These ideas, based on chemical potential, can often be extended to other chemical situations. One such situation, called Ostwald ripening, occurs in ice cream. Ice cream obviously contains many small ice crystals made of pure water. The chemical potential of a crystal of size r is increased by surface energy above that of a large crystal:

$$\mu_1(\text{water in crystal of radius}, r) = \mu_1^0(\text{water in large crystal}) + \frac{2\sigma}{r}, \qquad (5.4\text{--}4)$$

where the surface energy is described by the surface tension, σ. As always, the total free energy of the system will go to a minimum; thus, the small crystals will dissolve and the large crystals will grow. In ice cream, this is bad: ice cream with larger crystals won't taste as smooth. There is no way to eliminate this effect, but we can hope to slow it down. In ice cream, the retardation can come from the fat crystals also present in this frozen emulsion.

COLLOID STABILITY

These ideas of metastability, especially as applied to colloids, lead us to want to estimate how fast a metastable system will decay. The short answer is that the decay will normally be slow – months or years – unless the system becomes unstable. In these paragraphs, we want to discuss why this is so.

To do so, we need to return again to thermodynamics, and in particular to a normal phase diagram like that shown in Figure 5.4–1a). This diagram plots the temperature versus the average concentration in a binary mixture. At high temperature, the system forms one phase. At lower temperatures, below the equilibrium or "binodal curve," it will in principle form two phases, a saturated solution and a pure solid. At still lower temperatures, it will freeze into a physical mixture of frozen solvent and frozen solute.

In practice, however, this phase behavior can be more complicated. When the solution is cooled, it may not precipitate as it cools through the binodal curve. It may instead remain supersaturated for a long time, for at least hours or days.

The reason for this behavior is implicit in the discussion of Ostwald ripening above, and repeated in a different form in Figure 5.4–1b). When a solution is cooled below the binodal, it will have a lower energy if it precipitates and separates into two phases. To do so, however, it must first form small crystals, and these have higher energies than the solid, as the figure suggests. Even below the binodal, phase separation will occur only if some larger nuclei, perhaps dust, perhaps another solute, somehow circumvent this energy barrier and facilitate the formation of large particles of the lower energy phases.

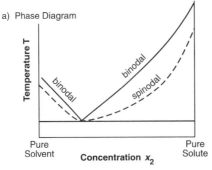

a) Phase Diagram

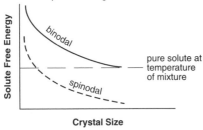

b) Binodal and Spinodal Energies

Figure 5.4–1. Binodal and Spinodal Lines in Crystallization. The equilibrium phase diagram, shown in a), has a metastable region between the binodal and spinodal. This region is the result of the increased surface energy of smaller crystals, as suggested in b).

Of course, this supersaturation cannot continue indefinitely. As the system continues to cool, the curve of free energy versus crystal size continues to drop, as suggested in the figure. Eventually, the curve drops so far that the crystal size required for phase separation becomes even smaller than that produced by random molecular fluctuations. Now, at this new "spinodal" condition, the system is truly unstable, and the system suddenly and catastrophically separates into two phases.

Thus the region between the first possible phase separation and the inevitable separation – between the binodal and the spinodal – is the region of metastability. For colloidal systems such as milk and paint, this same metastability exists, but it is expressed in the different terms shown in Figure 5.4–2. This figure plots the potential energy of two colloidal particles versus their distance of separation. When the particles are far apart, they don't interact much. When they are closer

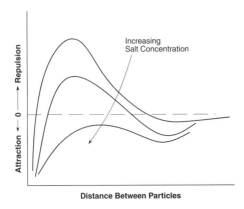

Figure 5.4–2. Stability of Colloidal Suspensions. As they get closer, colloidal particles are first attracted, then repelled, and finally irreversibly attracted. The repulsion can be reduced by adding salt.

together, particles of different electrical charge are attracted to each other. As the particles become still closer, they are repelled: they want to keep their proper distance. However, when they are forced even closer, they again attract, in almost a sort of shotgun marriage.

This progression of indifference to attraction to repulsion to fatal attraction can be changed, just as the tendency to phase separate may be altered. As just discussed, the tendency to phase separate is increased by cooling the system; the tendency to agglomerate is most easily increased by adding salts to the solution. This is because the behavior in Figure 5.4–2 is largely the result of electrostatics, and adding salt screens these electrostatic effects.

This colloidal destabilization is quantitatively described by the Derjaguin-Landau-Verwey-Overbeek theory, mercifully abbreviated as the DLVO theory. This theory predicts that within the metastable region,

$$\frac{[\text{salt concentration}][\text{counterion charge}]^6}{[\text{temperature}]^5} \leq \text{constant.} \tag{5.4–5}$$

Marvelously, the theory seems to work. Even the sixth-power dependence on the counterion charge, called the Schulze-Hardy rule, turns out to be verified. When the left-hand side of this equation is less than the constant, the system is metastable; when the left-hand side is greater than this constant, the colloid is unstable, and the colloidal particles agglomerate and phase separate. Indeed, Equation 5.4–5 can be used to explain why the ocean's salinity causes rivers to form deltas and why unsalted butter makes better Bearnaise sauce: in both cases, higher salt concentration can destabilize the colloid.

When this colloid destabilization occurs, the kinetics of colloid collapse is fast. To see why, we imagine a colloidal system originally containing N_0 particles per volume. Once destabilized, the particles agglomerate by colliding with each other, so that the concentration of particles drops with time. Commonly, this process is second order:

$$dN/dt = -kN^2, \tag{5.4–6}$$

where the rate constant has the dimensions of volume per time. This second-order rate equation, sometimes called a Smolukowski model, is easily integrated to give

$$1/N = 1/N_0 + kt. \tag{5.4–7}$$

Thus a plot of the reciprocal of the concentration should vary linearly with the time. In addition, the time $t_{1/2}$ when the number of particles is $(N_0/2)$, half the original number, is given by

$$t_{1/2} = 1/kN_0. \tag{5.4–8}$$

This time is a rough analog to the half-life used to describe radioactive decay.

The speed of colloid collapse now depends on the magnitude of the second-order rate constant, given by

$$k = 8\pi Dd, \tag{5.4–9}$$

where D is the diffusion coefficient of each colloidal particle and d is the particle's diameter. At first glance, we might expect that the rate constant will depend on the particle's size. Upon reflection, however, we realize that in the absence of electrostatic effects, which are suppressed by adding salt, the diffusion coefficient is given by the Stokes-Einstein equation:

$$D = k_B T / 3\pi\mu d, \tag{5.4–10}$$

where μ is the viscosity of the solvent. Thus the rate constant is just

$$k = 8 k_B T / 3\mu. \tag{5.4–11}$$

This rate constant implies rapid colloid aggregation after the colloid becomes unstable, as illustrated in an example at the end of this section.

RHEOLOGY AND MIXING

We next turn to the flow of microstructured products, and especially to the problems of their mixing. In these paragraphs, we are again reviewing topics in colloid chemistry that are often not stressed in the normal curriculum, but that are important for microstructured products. In the paragraphs above, we discussed thermodynamic stability. Now we want to give the same type of overview for flow.

When we push on a fluid, it will flow. When we push twice as hard, it normally will flow twice as fast. When we push very hard, the flow will become turbulent, and will vary approximately with the square root of how hard we push. This normal behavior, shown schematically in Figure 5.4–3, is called Newtonian flow, and occurs for simple fluids such as water, gasoline, and vegetable oil.

Such simple flows do not often occur in microstructured products. When force is first applied, there may be very little flow. Once a critical force occurs, flow may occur, but it often will increase more than linearly, as suggested in Figure 5.4–3.

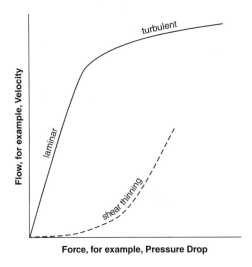

Figure 5.4–3. The Relationship Between Force and Flow. The solid curve represents Newtonian flow. The dashed curved describes flow of a shear thinning, non-Newtonian fluid.

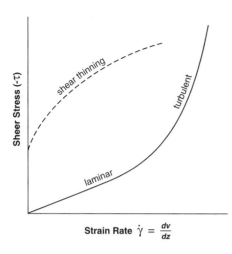

Figure 5.4–4. The Relationship Between Shear Stress and Shear Rate. The force in the previous figure is more exactly represented as a stress, i.e., a force per area. The velocity is better described as its gradient, on strain rate. As in the previous figure, the solid and dashed curves are for Newtonian and non-Newtonian fluids, respectively.

This faster increase is not the result of turbulence, but rather of non-Newtonian "shear thinning," where the product becomes less viscous as the flow gets faster. For microstructured products, true turbulent flow is rare.

This analysis of velocity versus flow is normally phrased in the more fundamental terms given in Figure 5.4–4. The velocity is not given directly, but as a shear rate: for a thin film, the shear rate is the velocity per film thickness. The force is given as the shear stress, τ, which is the force exerted on the thin film per unit area. The slope of this curve is the viscosity, μ:

$$\tau = -\mu(dv/dz). \tag{5.4-12}$$

For a Newtonian fluid in laminar flow, the viscosity is a constant. For a non-Newtonian fluid, it isn't; for the shear thinning fluid implied by the dotted line in Figure 5.4–4, the viscosity drops as the shear rate increases. Because the axes are reversed, the curves in Figure 5.4–4 are the mirror of those in Figure 5.4–3. Still, the science is the same; the focus is on how flow changes with force.

The non-Newtonian behavior of many microstructured products can offer very real advantages. For example, paint should be easy to spread, so it should have low apparent viscosity at the high shear caused by a paintbrush. At the same time, the paint should stick to the wall after it's brushed on, so it should have high apparent viscosity after it is applied. Many cleaning fluids and furniture waxes have should similar properties.

The causes of non-Newtonian flow depend on the colloid chemistry of the particular product. In the case of a water-based latex paint, the shear thinning is the result of breaking hydrogen bonds between the surfactants used to stabilize the latex. For many cleaners, the shear thinning behavior results from disruptions of liquid crystals formed within the products. The forces produced by these chemistries have consequences expressed over distances of micrometers. It is these forces that are responsible for the unusual and attractive properties of these microstructured products.

Although this non-Newtonian rheology may effect attractive product properties, it dramatically complicates mixing of these microstructured products. This is especially true at the larger scales required to produce commercial quantities. In these cases, our chief concern is to scale up the mixing, that is, to get the same degree of mixing at the production scale as we had in the laboratory.

How we carry out this mixing depends dramatically on whether we have laminar or turbulent flow. For high viscosity products, we will normally have laminar flow. In this case, we know of no general rule for scaling up the mixing. Our only suggested guideline is that the shear in any extruder die be duplicated.

For mixing low viscosity products, we will normally have turbulent flow, and we have more definite guidelines. In turbulent flow, the time for mixing, t_M, is approximately given by

$$t_M = l^2/4D, \tag{5.4–13}$$

where D is the diffusion coefficient of the species being mixed, and l is the size of a typical turbulent eddy. Because the diffusion coefficient doesn't change with the size of the mixer, we want to keep the eddy size the same. This eddy size may be shown to be

$$l \propto \left(\frac{\rho v^3}{P/V} \right)^{1/4} \tag{5.4–14}$$

where ρ is the product's density, v is its kinematic viscosity, and (P/V) is the power per volume in the mixer. The density and viscosity are constants; thus if the volume increases, the power must increase proportionally. We will illustrate these ideas in the examples that follow.

EXAMPLE 5.4–1   DESTABILIZING LATEX PAINT

A particular latex paint contains 20% by volume of 0.6-μm polymer particles. Smaller particles give a paint with more gloss; larger particles give a paint with more hiding power. When this paint is spread, the water in the emulsion evaporates, the colloid becomes unstable, and the particles fuse into a single smooth layer. In cases in which the paint is used at temperatures below the glass transition of the polymer, the polymer particles must be plasticized so that they fuse easily. This fusion is why a latex paint is hard to clean off after it dries, even though the original colloid is easily wiped up.

Paints like this are normally stabilized by surfactants, especially polyphosphates. In some cases, however, the surfactants can phase separate, and the colloid becomes unstable. Freezing can cause such an instability. If it becomes unstable, how long will it take for the paint to agglomerate?

SOLUTION

The easiest way to see how long the paint takes to agglomerate is to use Equation 5.4–8 to calculate the time to cut the particle concentration in half. To make this calculation, we need to find the original concentration, N_0, and the rate constant, k.

This concentration is

$$N_0 = \frac{\text{number of particles}}{\text{volume of paint}} = \left(\frac{\text{volume of particles}}{\text{volume of paint}}\right)\left(\frac{\text{number of particles}}{\text{volume of particles}}\right),$$

$$= \phi\frac{1}{[4/3\pi(d/2)^3]} = 0.2\frac{1}{\{4/3\pi[(0.6 \times 10^{-4}\,\text{cm})/2]^3\}} = 1.8 \times 10^{12}\,\text{cm}^{-3}.$$

From Equation 5.4–11, we can find the rate constant:

$$k = \frac{8k_B T}{3\mu} = \frac{8[1.38 \times 10^{-16}(\text{g cm}^2/\sec^2\,{}^\circ\text{K})]273{}^\circ\text{K}}{3(0.01\,\text{g/cm sec})} = 10 \times 10^{-12}\,\text{cm}^3/\sec.$$

Thus

$$t_{1/2} = \frac{1}{kN_0} = 0.05\,\text{sec}.$$

The number of particles is cut in half in 1/20 sec. Once the metastable colloid goes unstable, its collapse is catastrophic.

EXAMPLE 5.4–2   MAKING MORE ICE CREAM

A process for making ice cream, shown in Figure 5.4–5, begins by mixing the ingredients. After pasteurization, these ingredients are homogenized under high shear, and then cooled. The mixture is aged to allow fat crystals to form. It is

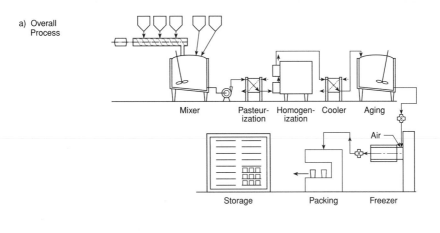

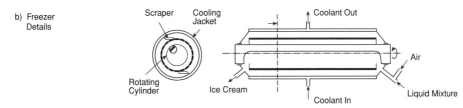

Figure 5.4–5. Schematic Diagram of a Process for Ice Cream Manufacture. As shown in a), the process begins with mixing, pasteurization, and cooling. Then, the ice cream is aged to allow metastable fat crystals to form. These stabilize the air added during freezing, and hence improve shelf life. Freezer details are given in b).

then aerated and pumped into the channel of an annularly shaped freezer. The outside of the freezer is chilled; the inside annular core rotates, scaping the outer surface with a helically shaped paddle. The extruded ice cream is then packaged and stored at low temperature to retard Ostwald ripening of the ice crystals.

You have been successfully operating a small ice cream plant, and are now thinking of building a new plant that is three times as large. How should you scale up the various mixing operations?

SOLUTION

To increase mixer volumes, and yet keep the same geometry, the diameter and heights of all tanks should increase $3^{1/3}$, or a factor of 1.44. The liquid mixture will have low viscosity, so in order to duplicate the original mixing, the homogenization and the aeration should use the same power per volume in the large tank as in the small one. You also want to use the same residence time in the new larger aerator as in the old one.

The difficulty may be the freezer. There, we want three times the surface area at the same shear rate as before. If we keep the residence time the same, the length must be unchanged, so the diameter of the freezer must triple. The power required in the larger freezer will also triple. The equipment cost for making ice cream will be reduced in the larger equipment, but not by as much as for most commodity chemical processes.

5.5   Device Manufacture

In this section, we review the manufacture of devices that produce chemical change. This section is not as long as the previous one because the devices involved are frequently small scale parallels to chemical process equipment. For example, an artificial kidney is a close analog of a packed tower for gas absorption, and a coffee maker is based on the same principles as ore leaching. The ideas for these processes are well covered in many chemical engineering courses, especially in those on unit operations. These ideas do not require review in this book.

However, there are some ideas often used in the design and manufacture of chemical devices that are not emphasized on conventional chemical engineering curricula. To identify these ideas, we consider the various products summarized in Table 5.5–1. As in the case of microstructured products, there are topics based in thermodynamics, especially chemical potential. There are concepts of mass transfer, central here, but often subordinated to heat transfer in conventional courses. There are ideas of catalysis, especially involving enzymes. These merit review here.

THERMODYNAMICS

We begin with two ideas based in chemical potential. As reviewed in the previous section, the chemical potential is nothing more than the average molecular energy (actually, the Gibbs free energy.) It is decreased by dissolution; it can be increased

TABLE 5.5–1 Typical Engineering Principles for Device Design

Intellectual Basis	Key Concept	Product
Thermodynamics	*Osmotic pressure*	Osmotic pump for drug delivery
	Electrochemical potential	pH electrode
	Heat of reaction	Wraps for sports injuries
Transport phenomena	Fluid flow	Infusion of physiological saline
	Film diffusion	Controlled drug release, slow release fertilizer
Unit operations	Heat transfer	Sleeping bags
	Mass transfer	Coffee maker, artificial kidney
	Heat and mass transfer	Home humidifier, heart-lung machine
Reaction engineering	*Enzyme reaction*	Glucose sensor, pregnancy test
	Nucleation	Smoke alarm

Note: The concepts that are italicized are those reviewed in this section.

or decreased by solute–solvent interactions; it can be increased by surface forces if the system is present as small drops or crystals. All this we discussed above.

The chemical potential can also be altered by electrical forces or by pressure. We discuss alteration by electrical forces first. The chemical potential μ_2 of an ion like chloride in aqueous solution is given by

$$\mu_2 = \mu_2^0 + RT \ln x_2 + z_2 \, \mathcal{F}\varphi, \qquad (5.5\text{–}1)$$

where μ_2^0 is a reference value, x_2 is the mole fraction of the ion, z_2 is its charge, $\mathcal{F}$ is Faraday's constant, and φ is the electrochemical potential acting on the ion. Note that these different terms can increase or decrease the chemical potential. The term containing x_2 decreases μ_2, because $RT \ln x_2$ is negative. The term containing φ can either increase or decrease μ_2.

Expressions such as Equation 5.5–1 are basic to the operation of many sensors, and especially to specific ion electrodes. To see why, imagine that we are measuring the potential in a solution with some sort of specific ion electrode. The selectivity of the electrode normally comes from surrounding the electrode with an ion selective membrane, through which only the ion of interest can penetrate. We are getting a reproducible signal with the electrode, but we are not at all sure if the electrode is really measuring the specific ion that we want to measure.

To test this, we make measurements with the sensor in two solutions whose mole fraction differs by an order of magnitude, that is, $x_2' = 10x_2$. The potential difference between our measurements is

$$\mathcal{F}(\varphi' - \varphi) = \frac{RT}{z_2} \ln\left(\frac{x_2}{x_2'}\right). \qquad (5.5\text{–}2)$$

If we are dealing with a monovalent cation, this potential difference is

$$\mathcal{F}(\varphi' - \varphi) = \left[\frac{(8.31\,\text{W sec/mol}\,°\text{K})298°\text{K}\,(\ln 0.1)}{(-1)96500\,(\text{amp sec/mol})}\right]\frac{\text{A} \times \text{V}}{\text{W}} = 0.059\,\text{V}.$$

$$(5.5\text{–}3)$$

Each decade change in concentration is worth 59 millivolts of potential. This variation, called the Nernst slope, is the key to testing many selective electrodes.

Like electrochemical potential, the ideas of osmotic pressure are also based on chemical potential. We consider a pure solvent separated from a solution by a membrane. The membrane is semipermeable, that is, it permits the passage of solvent, but not the passage of solute. As a result, solvent leaks through the membrane, diluting the solution. We can stop the solvent leak by putting the solution under pressure. Indeed, if we put the solution under high enough pressure, we can cause the solvent to flow out of the solution; that is, we can achieve reverse osmosis.

This effect can be quantified in terms of chemical potential. For simplicity, we consider the special case of no solvent flow, when the pressure applied to the solution stops the solvent flow, but doesn't reverse it. In this case, we can write for the solvent "1":

$$\mu_1(\text{pure solvent}) = \mu_1(\text{solution}),$$
$$\mu_1^0 = \mu_1^0 + RT \ln x_1 + \underline{V}_1 \Delta\Pi, \tag{5.5–4}$$

where μ_1^0 is a reference energy, the chemical potential of the pure solvent; $\underline{V}_1$ is the partial molar volume of the solvent; and $\Delta\Pi$ is the pressure applied to stop the solvent flow. Note that because the solvent mole fraction x_1 is less than one, the logrithmic term containing x_1 is negative, reducing the chemical potential. The pressure applied to the solution increases the chemical potential. Finally, note that this osmotic pressure term is the analogue of the electrostatic term in Equation 5.5–1: both reflect the effect of extra forces on the chemical potential.

In many cases, Equation 5.4–4 can be simplified. First, because the solution is dilute,

$$\ln x_1 = \ln(1 - x_2) \doteq -x_2 - \cdots, \tag{5.5–5}$$

where x_2 is the solute mole fraction. Second, the partial molar volume of the solvent, $\underline{V}_1$, which is the change in the volume per mole of solvent, is roughly equal to the reciprocal of the moles per volume, $1/c$, where c is the total concentration of solvent plus solute. Combining these relations, we have

$$\Delta\Pi = -\frac{RT}{\underline{V}_1} \ln x_1 \doteq c_2 RT. \tag{5.5–6}$$

This relation, called van't Hoff's law, was 80 years ago hailed as the analogy for liquid solutions of the ideal gas law. This analogy has since sunk in a morass of activity coefficients and other corrections. From our standpoint, we need only remember that osmotic pressure is proportional to concentration. We will return to this point later in this section.

ENZYME KINETICS

The last area that merits more emphasis for products than is usually given in chemical engineering courses is enzyme kinetics. Enzymes are proteins with catalytic activity. Proteins in turn are polymers with specific structures assembled from twenty amino acids. Because most proteins have their origin in living organisms, enzymes are normally effective catalysts only at modest temperatures, perhaps from 0 to 50°C. This is lower than the temperatures where normal petrochemical catalysts operate.

Enzyme catalysts are used in many sensors because they are exquisitely selective. For example, glucose oxidase reacts only with glucose, not with fructose; *l*-amino acid oxidase reacts only with *l*-amino acids. In many cases, the reaction produces a pH change, which is sensed by a pH indicating dye or by a pH electrode. This selectivity means that enzyme-based sensors can be extremely discriminating, which makes their kinetics important.

Enzyme catalysts often give reaction rates independent of the amount of reagent, which for enzymes is usually called the "substrate." This means that the amount reacted – that is, the signal produced – does not depend on the amount of substrate present. This is unusual. Usually, in chemical kinetics, we expect that the rate of reaction depends on the concentration of reagent. For a first-order reaction, doubling the concentration doubles the rate.

This zero-order reaction rate is a consequence of the mechanism by which many enzymes function. In this mechanism, the enzyme, E, and substrate, S, first form a complex, which then decays:

$$E + S \xrightarrow{K} ES \xrightarrow{k} P + E. \tag{5.5-7}$$

The complex formation often occurs rapidly, reaching an equilibrium described by the constant, K:

$$Kc_{ES} = c_E c_S. \tag{5.5-8}$$

The total amount of enzyme in its complexed and uncomplexed forms is constant:

$$c_{E0} = c_{ES} + c_E. \tag{5.5-9}$$

Finally, the decay of the complex is usually first order, so the rate is

$$r = kc_{ES} = \{[(kc_{E0})c_s]/(K + c_s)\}. \tag{5.5-10}$$

At high substrate concentration, the rate will obviously become independent of substrate concentration c_s. In fact, values of K, called the "Michaelis constant," are often small, perhaps 10^{-3} M. As a result, the rate of the enzyme reaction will be zero order at higher substrate concentrations.

EXAMPLE 5.5–1   AN ELECTRODE FOR MEASURING DODECYL SULFATE

An electrode for this anion consists of three basic parts: an inner filling solution, a liquid membrane, and the solution being tested. The inner solution is just 0.01 M of NaCl around an Ag/AgCl electrode. The liquid membrane consists of a chlorinated aromatic solution containing tributylhexadecylammonium chloride stabilized in a microporous support. The measurement also uses a second reference electrode that need not concern us here.

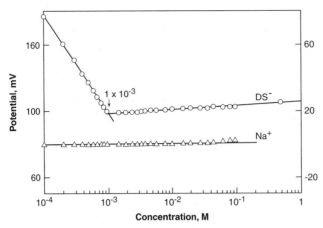

Figure 5.5–1. Electrochemical Potentials Measured for Sodium Dodecyl Sulfate with Added Sodium Chloride. The circles are the measured dodecyl sulfate's potentials. The triangles are the measured sodium potentials minus those found in the appropriate solutions containing only sodium chloride.

Data taken with this electrode are shown in Figure 5.5–1. Explain what these data mean.

SOLUTION

At low concentrations, the data are exactly what we should expect, showing a slope close to the Nernst limit of 59 mV per decade change in concentration. At a concentration close to 10^{-3} M, however, the data change abruptly, and the potential difference seems to become almost independent of concentration.

These data are consistent with micelle formation. At low concentrations, the dodecyl sulfate ions are unassociated and behave like other normal anions. Above a "critical micelle concentration" (cmc), in this case about 10^{-3} M, these anions associate into micelles, as discussed in the previous section. As more dodecyl sulfate anion is added, it does not go into solution, but into this aggregated phase. Thus the dilute data show that this detergent electrode is working, and the concentrated data give a detailed picture of micelle formation.

EXAMPLE 5.5–2   DESIGNING AN OSMOTIC PUMP

Drugs are often given orally, as pills taken every few hours. In many cases, this means that drug concentrations in the blood can fluctuate widely. Just after the pill is taken, the drug concentration jumps, sometimes briefly beyond the toxic limit. (In France, this is called "le burst effect.") After an hour or two, the drug concentration wanes, often dropping below the concentration where it is effective. Thus for many drugs, the blood concentration is occasionally too high and often too low, only periodically passing through the desired range.

These concentration variations have sparked many inventions aimed to provide a steady drug release. One such invention, shown in Figure 5.5–2, is called

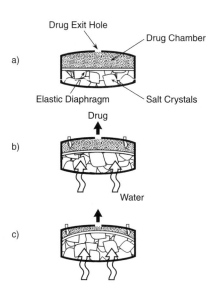

Drug Exit Hole

Drug Chamber

a)

Elastic Diaphragm   Salt Crystals

Drug

b)

Water

c)

Figure 5.5–2. An Osmotic Pump. This device has a rigid casing separated into two compartments by an elastic diaphragm. Water, pulled into the pump by osmosis, distends the diaphragm and steadily ejects the drug.

an "osmotic pump." The pump consists of a rigid housing capped with a semipermeable membrane. The housing is partially filled with a balloon, which in turn is filled with a solution of the drug whose delivery is to be controlled. The rest of the housing is filled with saturated brine in which sodium chloride crystals are suspended.

When this device is surgically implanted in the human body, the difference in osmotic pressure between the blood and the brine causes a water flow across the semipermeable membrane. The amount of the flow is proportional to the concentration difference across the membrane. The concentration difference of brine across the membrane stays constant because the brine inside the pump contains suspended salt crystals and these crystals dissolve as water flows in. Because the concentration difference stays constant, the flow is constant; because the flow is constant, the balloon is squeezed constantly to release drug solution. The beauty of this device is that the constant release of drug solution does not depend on the drug's properties, but only on those of brine and the membrane.

To design such a device, we plan to use a semipermeable membrane that has a reported permeability (defined as volume flux per unit pressure difference across the membrane) of 10^{-10} cm/sec kPa. What membrane area do we need to supply a release of 0.8 μL/hr?

SOLUTION

Saturated sodium chloride at a body temperature of 37°C has a concentration of about 5.4 mol/dm^3. Thus the osmotic pressure is

$$\Delta\Pi = c_1 RT = 2 \times \left(\frac{5.4\,\text{mol}}{10^{-3}\,\text{m}^3}\right) \times \frac{8.31\,\text{J}}{\text{mol}\,°\text{K}} \times 310°\text{K} = 28{,}000\,\text{kPa}.$$

The factor of two is the result of ionization. This pressure is high, almost 300 atm. The flux is

$$N_1 = \frac{1 \times 10^{-10}\,\text{cm}}{\text{sec kPa}}(28,000\,\text{kPa}) = 2.8 \times 10^{-6}\,\text{cm}^3/\text{cm}^2\,\text{sec}.$$

Thus the membrane area, A, is given by

$$\frac{0.8 \times 10^{-3}\,\text{cm}^3}{3600\,\text{sec}} = \frac{2.8 \times 10^{-6}\,\text{cm}^3}{\text{cm}^2\,\text{sec}} \times A;$$
$$A = 0.08\,\text{cm}^2.$$

This corresponds to a circular patch about 3 mm across.

5.6   Conclusions

This chapter begins the discussion of product manufacture, which is the last step in product design and development. This is the point where the study of chemical process design normally begins. In this book, we have gone back to earlier stages in the product development because chemists and chemical engineers are now frequently involved in these.

Key decisions on product manufacture involve intellectual property, supplying missing information, and setting final specifications. The questions of intellectual property center on the decision of whether or not to patent a product. For most chemical products, the answer is yes, for the exclusive license gained is worth the disclosure required. This is in contrast to many chemical processes, where we often seek a balance between patents and trade secrets. Supplying missing information is often demanded by the patent applications and helps get us ready for manufacture.

Setting final specifications is much more of a check on our thinking than a demand for creativity. We want to restate what we are doing, and what advantages we have over any competition. We are getting ready to invest a lot of money.

The rest of the material in this chapter is a review of chemical and engineering topics often important for chemical products. These topics, which include colloid chemistry and electrochemical sensors, are only superficially covered in most introductory courses. Their review here provides further support for the manufacture of products.

FURTHER READING

Arbuckle, W. S. (1996).   *Ice Cream*, 5th ed. Chapman & Hall, New York, ISBN 0412994917.
Edwards, M. F. (1998).   The Importance of Chemical Engineering in Delivering Products with Controlled Microstructure to Customers. Institute of Chemical Engineers, North Western Branch Papers No. 9.
Evans, D. F. and Wennerstrom, H. (1999).   *The Colloidal Domain*. Wiley, New York, ISBN 0471242470.
Kale, K. M., Cussler, E. L., and Evans, D. F. (1980).   Characterization of Micellar Solutions Using Surfactant Ion Electrodes. *J. Phys. Chem.* **84**, 593.

Koryta, J. (1991).   *Ions, Electrodes and Membranes*, 2nd ed. Wiley, New York, ISBN 0471930806.

Pressman, D. and Elias S. (eds.) (1999).   *Patent It Yourself*, 7th ed. Nolo Press, New York, ISBN 087337469X.

Sandler, S. I. (1999).   *Chemical and Engineering Thermodynamics*. Wiley, New York, ISBN 0471182109.

Sartori, G., Ho, W. S., Savage, D. W., Chludzinski, G. R., and Wiechert, S. (1987).   Sterically-Hindered Amines for Acid-Gas Absorption. *Sep. Purif. Meth.* **16**, 171–200.

Sefton, M., Hunkeler, D. J., Prokop, A., Cherrington, A. D., and Rajotte, R. (1999).   *Bio-artificial Organs*. New York Academy of Sciences, New York, ISBN 1573311944.

Stokes, R. J. and Evans, D. F. (1997).   *Interfacial Engineering: Where Physics, Chemistry, Biology and Technology Meet*, 2nd ed. Wiley, New York, ISBN 0471186473.

Warren, S. (1982).   *Organic Synthesis: The Disconnection Approach*. Wiley, Chichester, ISBN 0471101613.

6

Specialty Chemical Manufacture

As discussed in the previous chapter, chemical products can be conveniently separated into three categories: specialty chemicals, microstructured products, and devices for chemical change. Microstructured products and chemical devices were described in the Chapter 5. This leaves specialty chemicals, either pure compounds or chemical mixtures. Examples of these specialty chemicals include birth control pills, nonpolluting inks, and insect pheromones.

In this chapter, we discuss the manufacture of specialty chemicals. Such a discussion is part of the fourth step in product design. First, we identify the customer's need for a product. Second, we generate ideas that might fulfill this need. Third, we use chemical and engineering knowledge to select the most promising ideas. Fourth, we explore the manufacture of the product.

The manufacture of specialty chemicals is different from the manufacture of microstructured products. The properties of microstructured products usually depend on physics and physical chemistry, and not on chemical reactions. For example, the "smoothness" of ice cream depends on the size of air bubbles and ice crystals in the product. In the same sense, the manufacture of chemical devices focuses on device architecture, on how the components of the final product are assembled. We need schematic drawings of the product, which can be clustered into different elements. We need to think through how the different elements are assembled. But we are not concerned with chemical reactions as part of this assembly.

The manufacture of specialty chemicals also involves architecture, but of molecules, not the macroscopic devices discussed in the previous chapter. By this point, the chemical synthesis – how the raw materials react to make the product – will be known. This synthesis will have been developed by research chemists. Our first step, outlined in Section 6.1, is to collect and verify the chemists' results.

The second step, outlined in Section 6.2, is to develop the reaction engineering needed to manufacture the chemical. We know from the chemist how the different chemical groups are assembled to make the target molecule. We are concerned with the chemical kinetics, for these describe how much product we can make in a given time. This reaction engineering will also emphasize how we add our reagents, for this will affect the selectivity of our synthesis.

Once the reaction engineering is established, our concern will shift to the separation and purification of our chemical product. As Section 6.3 shows, this separation can dominate the product's final cost. Earlier chemical studies are often of limited value, so this is one place in product design where chemical engineering can be central to a successful chemical product.

Finally, in Section 6.4, we turn to issues of process scale-up. The scale-up of specialty chemical processes differs from that normally discussed in chemical engineering because specialties use generic, batch equipment, suitable for many different chemical products. Nonetheless, the scale-up of specialty chemical processes will depend on dimensionless groups, the greatest intellectual concept that comes from engineering. By applying dimensionless groups with ideas of reaction engineering and separation processes, we can achieve the successful manufacture of specialty chemicals.

6.1   First Steps Toward Production

We begin by defining what we mean by specialty chemicals, and by emphasizing how these specialties differ from the commodity chemicals implied by most studies of chemical engineering. There are two big differences. First, specialty chemicals tend to have high added value. In other words, the specialty product will sell for much more than the cost of the raw materials. A good example is an antibiotic, selling for perhaps $10/kg, but made by fermentation from agricultural wastes costing pennies per kilogram. In contrast, a commodity chemical such as toluene, selling for perhaps $0.20/kg, is made from alkanes costing around $0.15/kg.

The second major difference between commodity and specialty chemicals is the amount produced. Specialty chemicals are usually made in quantities totalling less than 10^6 kg/yr. In contrast, commodities are made in quantities greater than 10^7 kg/yr.

As a result, specialty chemicals are usually made in reactors not dedicated to a specific product but useful for many different products. The reactors are batch, often not run 24 hours per day. Thus, the only significant kinetic question may be whether the reaction can be finished in an 8-hour shift. The reactors are not built to optimize production of a given chemical, but to be flexible enough to produce different materials in different residence times. The key is not the optimization of one product, but the scheduling of many products.

These characteristics of specialty chemicals mean that their manufacture is often different to that expected by chemical engineers. For example, most chemical engineers expect to translate the batch reactions that chemists study in stirred, round-bottom flasks into continuous reactions in tubular plug flow reactors. They expect this translation because they will get more conversion per reactor volume, and hence more product per investment in capital equipment. However, the optimal conversion per reactor volume is harder to achieve for each of perhaps twenty different reactions in the same reactor. In this case, most engineers will wind up using stirred stainless steel tanks that mimic the chemists' round-bottom flasks.

The engineer's task is often to scale up the chemist's results from a few grams to a few kilograms. For the case of pharmaceuticals, the scale-up will often be wanted as quickly as possible, for the few kilograms will be used for the new drug's clinical trials. These clinical trials, first on isolated cells, then on animals, and finally on humans, are likely to limit the rate of product development. The need to get these trials started is the reason that the engineer will be pushed to make the scale-up as rapid as possible.

However, there is a major risk in the rapid scale-up of pharmaceuticals. The Food and Drug Administration (FDA) in the USA insists that the drug produced commercially must be made by the same process as that used to make the drug for the clinical trials. For example, if you use methylene chloride as a solvent in your original scale-up, you must use it in your commercial process. You cannot arbitrarily replace methylene chloride with toluene, even if you later discover that it gives you a higher yield of a purer product. If you want to make this replacement, you must first petition the FDA to do so, a slow and sometimes seemingly capricious procedure. Thus, although you want to do the scale-up fast, you must remember that you must later live with your process choices.

In the rest of this section, we discuss in more detail the manufacture of chemical products, including pharmaceuticals. This product manufacture is best organized under three headings: extension and verification of the laboratory work, engineering the reactions, and separation processes. These topics are explored in the following paragraphs.

EXTENDING LABORATORY RESULTS

In many cases, a chemical product will have been selected for manufacture after only small amounts have been made in the laboratory. In the case of drugs, these small amounts will have been tested for pharmacological activity, usually against target microorganisms or a few living cells from a hybridoma. These sketchy tests are often a good indication of whether the drug is effective.

If the product is effective, we will want to make more of it. In order to achieve this, the chemist who made the initial material must work with the process engineer whose responsibility is to make much more. Such cooperative enterprise is often fractious, filled with unintended and intended insults in both directions. To illustrate this, consider the following conversation to make a steroid, like that in birth control pills:

Chemist: This is an easy reaction which anyone intelligent should be able to run. I just dissolve the crude steroid in methylene chloride and then add n-butyl lithium. The reaction is . . . Wait, let me put it in terms you'll understand. At $-40°$C,

$$A + B \rightarrow AB.$$

You can't run too long because there's a side reaction:

$$AB + B \rightarrow AB_2.$$

I then add acetone, which knocks out the product (i.e., causes it to precipitate). I decant the solvents and add DMF (dimethylformamide) to redissolve it. Then I add water to make the alcohol:

$$AB + H_2O \rightarrow AOH + BH.$$

All these reactions are pretty exothermic. Still, they run easily, though the overall selectivity is often low, around 40%. You shouldn't have any trouble getting that higher.

Engineer: Why is the selectivity so low?

Chemist: I don't know. It often is in reactions like these.

Engineer: How much does the temperature increase?

Chemist: Quite a lot. Even at $-40°C$, you can see the temperature jump when you add the *n*-butyl lithium. However, I've kept the temperature rise small by running in an acetone-CO_2 bath. Sometimes, I've kept it from jumping too much by turning off the stirrer for a while.

Engineer: Can you use any different solvents?

Chemist: I don't know. You probably can't replace methylene chloride; it really is the best for these reactions.

Engineer: You remember that it's viewed as a dangerous carcinogen.

Chemist: Yeah, but lots of chemicals are dangerous.

Engineer: Could methylene chloride be replaced with butyl acetate?

Chemist: I don't know. Look: I really like methylene chloride. It works really well and I think you'll have trouble replacing it.

Engineer: Did you ever check for the maximum temperature rise in this reaction?

Chemist: No, but it could be big, enough to boil the solvent. But you can slow the reaction by shutting down the stirring.

Engineer: Does that work if the reaction mixture starts to boil?

Chemist: I don't know. My experiments never boiled.

Engineer: Why do you always run in a round-bottom flask? You could get faster conversion in a tubular reactor.

Chemist: Look, I need to slow the reaction down, not speed it up. When it runs too fast, it makes too much by-product. Then the product goes brown, not white, like it probably should be.

Engineer: How can you remove the color?

Chemist: I don't know. Sometimes activated carbon works on problems like these.

Engineer: Can you try to get any purification when you make the acetone knock-out?

Chemist: You mean add the acetone slowly so that you get purer crystals? That's a good direction to go, though it's hard at $-40°C$. I didn't do it, because I was just trying to rough out the process chemistry.

Engineer: Did you measure the purity of that intermediate precipitate?

Chemist: No. I don't think it is that important.

Engineer: How did you separate the product? The one after hydrolysis.

Chemist: Actually I didn't. I just ran the solids that were knocked out and hydrolyzed through the HPLC (high pressure liquid chromatograph). I knew where the peaks should be because of earlier experiments using combinatorial chemistry.

Engineer: Do you know how to purify the product?

Chemist: Sure.

Engineer: I mean at large scale.

Chemist: But that's your job. I finished this one, and I did it right. I've got other reactions to run. Come back and see me if you need help. This isn't hard. See you later.

This ended the discussion.

What should the engineer have learned? He must scale up a highly exothermic reaction whose selectivity is strongly temperature dependent. The reaction is possibly mass transfer controlled, because its rate depends on stirring. The separation of the reaction products will include raw materials and the results of side reactions. Separation by adsorption – the basis of chromatography – works, at least on a small scale. Solvents are important, but largely uninvestigated.

In a case like this, we must first check the chemist's results. We must repeat the reactions in round-bottom flasks, carefully watching the temperature versus time. We should imitate the way that the chemist combines the reagents, and we should use the same solvents, even the methylene chloride. We should separate the products by HPLC. In most cases, we will not initially get results that are as good as the chemist's, a result of the chemist's greater skill and of the inadvertent omission of nuances of chemical technique. Eventually, we should equal or surpass the chemist's laboratory results. We are then ready for the reaction engineering.

REACTION ENGINEERING

We have duplicated the chemist's results; we need to consider the speed and the selectivity of the chemical reaction. The chemist has shown that this synthetic route is possible. We need to discover how much it can make.

We normally begin by seeking the rate-limiting steps of the various reactions. In the example mentioned above, we will probably measure the concentration of the limiting reagent versus time. In many cases, the limiting reagent will be the most expensive material, and the excess reagents will be cheaper. However, in some cases, including this one, we may use the expensive reagent in excess and make the limiting reagent the less expensive one, n-butyl lithium in this example. We will do so to minimize the side reaction, which involves reaction of two n-butyl lithiums with one basic molecule.

Determining the rate-limiting step is detailed in texts on chemical kinetics, so we give only a synopsis here. Normally, we will want to know the effect of changing the concentration of the limiting reagent. We will cut that concentration in half, and measure how the reaction rate changes. If the rate also halves, the reaction is first order; if the rate drops four times, the reaction is second order; if the rate doesn't change, the reaction is zeroth order.

We then measure how the rate changes with temperature and with stirring. We can make a first guess at the rate-limiting step by using Table 6.1–1. Knowing this limiting step will supply a key to reactor scale-up, reviewed in Section 6.3.

EXAMPLE 6.1–1   PENICILLIN MODIFICATION

The addition of a phenyl group to a β-lactam ring shows the following kinetics:

time, sec	concentration, β-lactam
0	0.0110 M
200	0.0072
400	0.0047
600	0.0032

TABLE 6.1–1   Identifying the Rate-Limiting Step

Limiting Reagent	Variation with Temperature	Stirring	Rate-Limiting Step	Remarks
First order	Strong	Weak	Chemical kinetics	Most common case
	Weak	Strong	Mass transfer	Common for pharmaceuticals
Second order	Strong	Weak	Chemical kinetics	Uncommon; if one reagent is in excess, becomes first order
Zero order	Varies	Weak	Chemical kinetics	Often indicates catalysis

Note: This brief summary should be used as an introduction, supplemented by books on reaction engineering.

If the β-lactam is the limiting reagent, what is the reaction order? What is the rate constant?

SOLUTION

A plot of the logarithm of β-lactam concentration versus time is nearly linear. This is characteristic of a first order reaction. To show why, consider the mass balance

[rate of change of β-lactam concentration] = [reaction rate of β-lactam],
$$V(dc_1/dt) = -kc_1 V,$$

where V is the volume of the reactor, c_1 is the β-lactam concentration, t is the time, and k is a first order rate constant. If the initial concentration of the β-lactam is c_{10}, this equation is easily integrated:

$$c_1/c_{10} = c^{-kt}.$$

Thus a plot of $ln\ c_1$ versus t should have a slope of $-k$. In this case, k equals about 2×10^{-3} sec^{-1}.

EXAMPLE 6.1–2   ETCHING A PHOTORESIST

Imagine that we are etching a silicon wafer coated with a new optically sensitive photoresist with a dilute solution of aqueous sodium hydroxide. The reaction rate shows an activation energy around 30 kJ/mol, suggesting the reaction may be strongly influenced by chemical kinetics. However, the reaction rate in 0.16 M of NaOH also depends on the spinning speed of the wafer, as shown by the isothermal data below.

wafer rotation, rpm	rate constant, sec^{-1}
6	0.53
9	0.61, 0.61
15	0.70, 0.66
30	0.81, 0.79, 0.88
70	1.16

What is the reaction mechanism?

SOLUTION

In many cases, including this one, the rate is affected both by chemical kinetics and by mass transfer. In cases such as these, we may show that the overall reaction rate constant, k, is given by

$$\frac{1}{k} = (1/k_{surface}a) + (1/k_D a),$$

where $k_{surface}$ is the rate constant for surface reaction, k_D is that for diffusion (i.e., the mass transfer coefficient), and a is the surface area per volume. Because these two reactions occur sequentially, this result is often said to correspond to two chemical resistances in series, and it is compared to Ohm's law of electrical resistances in series.

At constant temperature, we expect $k_{surface}$ to be a constant, but k_D to vary with stirring. In most cases, k_D varies with stirring speed to the power of 0.3 to 0.8. The most common power is 0.5, which is that for the dissolution as given above. Thus we can expect

$$\frac{1}{k} = (1/k_{surface}a) + (B/\omega^{1/2}),$$

where ω is the speed of rotation and B is a constant. Thus a graph of $(1/k)$ versus $(1/\omega^{1/2})$ should be linear, with an intercept proportional to the chemical rate constant, and a slope related to the diffusion constant. Such a graph, sometimes called a Wilson plot, does work in this case, as shown in Figure 6.1–1.

6.2   Separations

We now turn from the problems of making our target chemical to those associated with separating and purifying it. These problems are often complicated by the tendency of specialty chemicals to be produced at high dilution. For example, therapeutic proteins may be synthesized by genetically modified mammalian cells at concentrations around 20 μg/L. We need to produce pure protein from this solution, a task analogous to finding a hundred or so persons who have a single uncommon genetic defect in the entire world's population.

Separations of mixtures of dilute chemicals usually involve two groups of problems. First, we must plan in what sequence we intend to separate the various compounds in our reacted mixture. This planning can be facilitated by a few heuristics, rules of thumb that have often proved reliable in the past. Second, we should review the types of separation processes that are most likely to be useful for

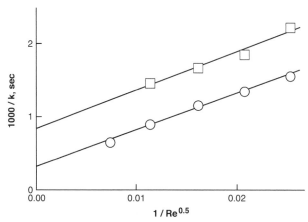

Figure 6.1–1. Dissolution Rates for a Light-Sensitive Photo-resist. These experimentally determined rates in two different etchants are analyzed as an overall mass transfer coefficient k, which is a function of both chemistry and diffusion. In this Wilson plot, values of k^{-1} are plotted vs. the reciprocal of velocity, shown as a square root of Reynolds number $\mathrm{Re}(=dv/\nu)$. The intercept on the graph is the chemical contribution; the slope is related to the effect of diffusion.

these products. Attractive processes usually do not include distillation, which is a major difference from commodity chemicals. These two groups of problems are discussed in the following paragraphs.

HEURISTICS FOR SEPARATIONS

Separating specialty chemicals will normally begin with the contents of a batch reactor. These contents will be fed to generic separation equipment to produce perhaps 10–100 kg of product. Although the separations involved vary widely, the following heuristics can guide how we proceed. These are given in their approximate order of importance:

1. Concentrate the product before purifying.
2. Remove the most plentiful products early.
3. Do the hardest separations last.
4. Remove any hazardous materials early.
5. Avoid adding new species during the separation. If you must add them, remove them promptly.
6. Try to avoid extreme temperatures by using different solvents.

These important guidelines merit additional discussion.

1. Concentrate Before Purification

This heuristic suggests that the first step in any separation train should focus simply on taking the dilute feed and concentrating both the product and the principal

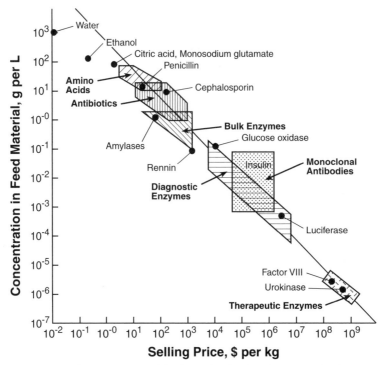

Figure 6.2–1. Sherwood Plot of Selling Price Against Concentration in Feed Material. Although this particular plot is for biological products, similar graphs can be produced for other classes of materials.

impurities. This most often means removing a lot of water or solvent. At this stage, we should not worry about any selectivity; we just want to reduce the volumes we are processing.

This heuristic gains considerable support from a graph of product concentration in the feed versus product selling price. Such a graph, shown in Figure 6.2–1, is sometimes called a "Sherwood plot," after the eminent chemical engineer, Thomas Sherwood. This logarithmic plot covers twelve orders of magnitude, so that it is presumably more accurate than many log-log plots. It includes chemicals such as water and ethanol, as well as therapeutic enzymes costing 10^{12} times more. Mainstream antibiotics such as penicillin fall roughly in the middle of this range.

The implication of Figure 6.2–1 is that concentrating the product is more important than separating it. To see why this may be true, we first note that the slope of the line in Figure 6.2–1 is minus one. Thus, if the product's concentration in the feed can be improved ten times, then the cost should drop ten times.

As an alternative test of this implication, imagine that for a particular antibiotic we have developed a highly selective separation. This new separation is inexpensive as well. Then this antibiotic should appear as a point well below the line in Figure 6.2–1. Like other antibiotics, it would have a low feed concentration; but, unlike the others, it could be cheaply separated and so have a low price. The fact that no points well off the line occur in the figure suggests that whether a selective separation exists will not matter as much as the original concentration in which the antibiotic was manufactured.

We recognize that these conclusions may be flawed, that in some cases we may do better to separate selectively before we concentrate. This could cut the cost of making our product in half, and still not make a dramatic difference to Figure 6.2–1. Still, the message in most cases is clear: concentrate first.

2. Separate the More Plentiful Products Early
3. Do the Difficult Separations Last
4. Remove any Hazardous Materials Early
These three heuristics, often given for any chemical separations, are best understood by imagining we want to sort the tableware removed from a big dishwasher. The tableware has all sorts of utensils: forks, spoons, spatulas, knives, and so forth. We want to sort them and put them away. We will first remove any sharp knives, and then we will quickly sort the forks and spoons. Finally, we will separate Aunt Evetta's silver coffee spoons, used only on special occasions, from the other tea spoons that we use every day.

What we have done is an illustration of these three heuristics. We separate the sharp knives first because they can cut us; they are a potential hazard. We sort the forks and spoons early, because there are a lot of each and because the separation is easy. We separate Aunt Evetta's spoons last because they look similar to some of the other spoons.

These heuristics are certainly valuable for commodity separations, where they are widely useful. We believe that heuristics 3 and 4 are also applicable in specialty chemicals, but we are less sure that heuristic 2 is as reliable. The difference is a reflection of the value of by-products. For commodities, we will seek markets for any by-products. For example, if we are making ethylene, we will certainly find uses for any by-product hydrogen. However, for specialties, we may not have any by-products that have significant value. In these cases, heuristic 2 will not be as useful.

5. Avoid Adding New Species. If You Must Add Them,
Remove Them Promptly
We will need to add new species in many specialty chemical separations. We will add solvents to extract many fine chemicals from the original extraction mixture. We will use adsorbants, especially ion exchangers, for purification. We will add detergents to lyse cell walls and hence release precipitated proteins – inclusion bodies – that are of therapeutic value. We do so because adding these species gives us a more effective separation.

However, the caution that we remove these added species quickly is the real message of the heuristic. In some cases, we will do so almost automatically. For example, we will normally evaporate an extraction solvent or further concentrate our product back into water at a different pH. But in some cases, these added species will be much harder to remove. The detergents used to lyse cells are an

TABLE 6.2–1   Key Separation Processes for Specialty Chemical Products

Class 1 Distillation	Class 2 Workhorses Requiring Thought	Class 3 Other Important Processes

Fractional distillation
Steam distillation
	Extraction
	Adsorption
	   (including ion exchange)
	Crystallization
	   (including precipitation)
		Drying/evaporation
		Filtration
		   (including ultrafiltration)
		Centrifugation
		Absorption
		Membrane separation
		Electrophoresis

excellent example. The detergents that rupture cell walls most effectively are also those which just keep moving on through the entire separation process until they are hardest to get out of the final product.

6. Try to Avoid Extreme Temperatures by Using Different Solvents
High temperatures can decompose many specialty chemicals, and low temperatures can be expensive. After all, we are not going to want to put our filter press, used for several different products, into a walk-in freezer. In many cases, we can avoid these challenges by switching solvents. In doing so, we will have trouble getting help from the synthetic chemists, who will normally have identified a few that work well and will not be sympathetic with our efforts to change reaction conditions in order to make the purification easier. In addition, we must remember that our choice of solvents is more binding than normal, for it may violate the manufacturing procedure approved by the FDA. Although we may get some help from the solubility parameters in Section 4.1, we are going to depend mostly on new experiments.

THE MOST USEFUL SEPARATIONS

We now want to look at the separation processes themselves. Many good books on separations give wonderfully elaborate lists of these processes, cleverly organized around the chemistry and physics responsible for their effectiveness. Distillation is based on phase equilibrium; filtration is based on size differences; and membrane separations are based on diffusion rates. Such an organization makes sense for these books.

Here, our objective is different. We want to discuss which separations are the most valuable for specialty chemicals and what the key features are for these important processes. We want to list some basic ideas and some key results, but we do not want to duplicate what is detailed in those good books on separations. We want to provide a primer on how to plan the separation train.

We begin in Table 6.2–1 with a list of the most important separations, broken into three classes. Distillation is literally in a class by itself. Fractional distillation is usually not important for specialty chemicals because these tend to have low volatility and to be thermally unstable. However, it is the most important separation process for commodity chemicals. This process is basic to the estimated 40,000 columns that consume about 6% of the energy used in the USA. Fractional distillation will be important to many peripheral steps in specialty manufacture, including the recycle and reuse of solvents. However, because fractional distillation is not often central to separating specialty products from reaction mixtures, it is not reviewed in detail in this book.

One method of distillation that is quite common for specialty chemicals is steam distillation. In this method a temperature-sensitive organic that is only partially miscible in water is distilled from a two-phase liquid mixture. The presence of the aqueous rich phase lowers the boiling point and so allows distillation of the organic without its decomposition. Steam distillation is the most common method of obtaining extracts from plants; essential oils are an everyday example.

The thermodynamics of steam distillation does merit review. In a two component, single-phase liquid mixture, a first approximation to the total vapor pressure is given by Raoult's law:

$$p = x_1 p_1 + x_2 p_2, \tag{6.2-1}$$

where x_i is the mole fraction of component i in the liquid mixture and p_i is the vapor pressure of pure i at the temperature being considered. The vapor pressure is always lower than that of the more volatile component. However, when the two components are immiscible and a two-phase liquid mixture forms, the vapor pressures are approximately additive:

$$p = p_1 + p_2. \tag{6.2-2}$$

The vapor pressure is now always higher than that of the more volatile component on its own. Because the boiling point is the temperature at which the total vapor pressure equals the external applied pressure, this implies that the boiling point of the two-phase mixture is lower than both the boiling points of the individual components. Thus even relatively nonvolatile species can be steam distilled at a temperature lower than the boiling point of water. Moreover, once the distillate is condensed, the two liquids separate out again and so the removal of the added water is easy.

The second class of separation processes in Table 6.2-1 includes extraction, adsorption, and crystallization. These processes tend to be the most important for specialty chemicals. All can be used to concentrate dilute solutions, and so are key to the final product price. Most are capable of achieving selectivity, although this may require careful staging. All three require above average amounts of energy – often as free energy – in order to effect the product's purification. Each merits the additional discussion given below.

The separations in the third class of Table 6.2-1 are important to specialty chemical separation but rarely cause major problems in the separation. The best example is drying. We will often collect a final product as crystals, wet with the mother liquor from which they have been crystallized. We will dry these crystals by warming them, evaporating the solvent from the occluded liquor. This is an important step, but it is not hard. It will rarely be central to the success or failure of our process. In the same sense, absorption is the key to treating "acid gases," streams containing CO_2, H_2S, SO_2, and NO_x that occur so often in petrochemical processes. Similar streams are much rarer in specialty chemical manufacture. As a result, we will discuss key aspects of absorption and the other separations in this third class only tangentially.

We now return to the big three: extraction, adsorption, and crystallization.

Extraction

In extraction, we normally begin with the dilute solution containing the specialty product produced by our batch reactor. We contact the solution with samples of solvent in which the product is more soluble. We thus concentrate the product, not necessarily selectively.

Three aspects of this process merit discussion: how we choose the solvent; how much separation we can get in one batch; and how we can purify by using repeated extractions. One easy way to choose the extraction solvent is to look at the solubility of the product in the original solution relative to that in the extraction solvent. If the relative solubility is small, then that extractant is a good choice.

To justify this criterion for selecting solvents, we remember that for the dilute product "1" in equilibrium between feed and solvent,

$$\mu_1(\text{in solvent}) = \mu_1(\text{in feed}), \tag{6.2–3}$$

$$\mu_1^0(\text{in solvent}) + RT\ln cx_1(\text{in solvent}) = \mu_1^0(\text{in feed}) + RT\ln cy_1(\text{in feed}), \tag{6.2–4}$$

where μ_1 is the chemical potential, μ_1^0 is its value in a standard state (defined here as the chemical potential at unity concentration in the solvent or feed), c is the total concentration, and x_1 and y_1 are the mole fractions in the solvent and feed, respectively. In writing these equations, we are assuming ideal mixing, but the chemical potentials are relative to different standard states for the solvent and feed. We now define the partition coefficient, m, as

$$m = \frac{y_1(\text{in feed})}{x_1(\text{in solvent})},$$
$$= \exp\left[\frac{\mu_1^0(\text{in solvent}) - \mu_1^0(\text{in feed})}{RT}\right]. \tag{6.2–5}$$

We want a small value of m for an efficient extraction.

To find this m we need to estimate the exponential mess in this equation. We do so by recognizing that at saturation in the feed,

$$\mu_1(\text{as solid}) = \mu_1^0(\text{in feed}) + RT\ln cy_1(\text{saturated in feed}). \tag{6.2–6}$$

Similarly, at saturation in the solvent,

$$\mu_1(\text{as solid}) = \mu_1^0(\text{in solvent}) + RT\ln cx_1(\text{saturated in solvent}). \tag{6.2–7}$$

Solving for the standard state chemical potentials and inserting the results into the equation for the partition coefficient, we find

$$m = \frac{y_1(\text{in feed})}{x(\text{in solvent})} = \frac{y_1(\text{saturated in feed})}{x_1(\text{saturated in solvent})}. \tag{6.2-8}$$

To find the low m that we seek for efficient extraction, we should look for a low relative solubility.

The second key to extraction is to estimate how much separation we can get in a single batch, that is, in a single stage. We start with a feed of solute concentration, y_{10}, and amount, H, and a pure solvent with solute concentration, $x_1 = 0$, and amount, L. Then from a mass balance, we find

$$H y_{10} = H y_1 + L x_1. \tag{6.2-9}$$

Combining with the partition coefficient, m, we find the fraction extracted, f, is given by

$$f = \frac{L x_1}{H y_{10}} = \frac{1}{[1 + (mH/L)]}. \tag{6.2-10}$$

Again, note that a small value of m will give a large fraction extracted. Although these relations are accurate only for dilute solutions, they are basic to understanding.

Before we proceed, we should remember that this batch extraction can be solved graphically. There are two equations:

$$y_1 = y_{10} - \frac{L}{H} x_1, \tag{6.2-11}$$

$$y_1 = m x_1. \tag{6.2-12}$$

The first is the mass balance; the second is the thermodynamic equilibrium. We can clearly solve these equations analytically; we did so in the preceding paragraph. We can also solve them graphically. The first equation begins on the ordinate at y_{10}, drops with a slope of $(-L/H)$, and crosses the abcissa at $(y_{10}H/L)$. The second begins at the origin and rises with a slope of $(+m)$. They intersect at $[y_{10}/(m + L/H)]$, $[y_{10}/(1 + L/mH)]$, which is of course the analytical solution to the problem. Try this: it works.

The reason that this graphical solution is important is that it also works when the mass balance and thermodynamic equilibrium are not simple functions. This can occur for concentrated extractions, for partly miscible solvents, and for most adsorptions. Just remember which lines are mass balances and which are thermodynamic equilibria, and your understanding will jump dramatically. We now return to our basic problem of concentrating the product by extraction.

In most cases, we will use repeated extractions to recover our product from the initial feed taken from the reactor. In many cases, we will split the solvent into n equal portions, and wash the feed n times with these portions. The total fraction,

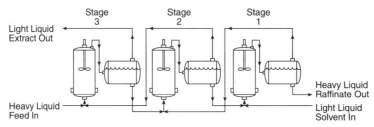

Figure 6.2–2. A Three-Stage Cascade of Mixer-Settlers Used to Purify by Extraction. Each mixer-settler unit is treated as a separate stage. Note where the heavy feed and light solvent enter the cascade.

f, extracted in n extractions can be shown to be

$$f = \frac{\sum_{i=1}^{n}(L/n)x_{1i}}{Hy_{10}} = \frac{1}{[1 + (mHn/L)]}\sum_{j=1}^{n}\left[\frac{1}{(L/mHn) + 1}\right]^{n-1}. \qquad (6.2\text{–}13)$$

After these repeated extractions, we will normally pool the extracts and evaporate much of the solvent, under vacuum if we need to keep the temperature low.

In addition to using extraction for product recovery and concentration, we will sometimes use extraction as a means of purification. This commonly makes use of mixer-settlers, arranged as stages. A typical mixer-settler, shown in Figure 6.2–2, consists of one stirred tank, called the mixer, into which both phases are pumped. The resulting emulsion of the feed and solvent phases is then pumped to a second unstirred tank, called the settler, where the two phases are allowed to separate.

When used for purification, the mixer-settlers are often arranged in a staged cascade, also shown in Figure 6.2–2. Now each box represents a mixer-settler pair. For simplicity, we have assumed an aqueous feed and an organic solvent. Stages 2 and 3 concentrate the product, removing it from the aqueous stream into the organic extract. Stage 1, however, dilutes the product because of washing the organic with the pure water, the second aqueous feed to this cascade. However, although Stage 1 dilutes the product, it also purifies it, preferentially washing away impurities. Thus extraction can purify as well as concentrate. The price of this purification is dilution.

Adsorption

The second of the three key separation processes is adsorption. In adsorption, we normally contact a feed containing the product with a solid adsorbent. Because the adsorbent is usually microporous, it has a large surface area on which it can adsorb the product. These solute-surface interactions are frequently more selective than the solute-solvent interactions that occur in extraction. Thus adsorption is especially effective for product purification, though it can also be used for product concentration.

Like extraction, adsorption is conveniently discussed as three topics: how we choose the adsorbent, how it will work in batch, and how we will use it to purify the product. The choice of the adsorbent depends on experimental measurements

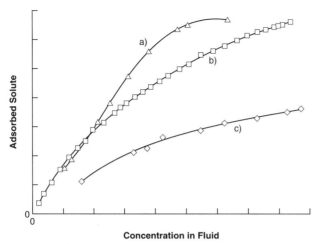

Figure 6.2–3. Isotherms for Various Adsorbents. The data in
a) are for water on silica gel (W. L. McCabe, J. C. Smith, and
P. Harriott, *Unit Operations in Chemical Engineering*; McGraw-
Hill, New York, 1996, p. 691); those in b) are for carbon dioxide on carbon (E. Weyde and E. Wicke, *Kolloid* Z. **90**, 156,
1940); and those in c) are for bovin serum albumin on sepharose
(S. Yamamoto et al., *Ion Exchange Chromatography of Proteins*,
Dekker, New York, 1988, p. 119). Note all these isotherms are
favorable because they all bend downward.

of the equilibrium between product adsorbed versus product in solution. These experimental results, called isotherms, are often presented graphically, as in
Figure 6.2–3. Isotherms are often nonlinear, implying that the thermodynamics is
more complex than that responsible for the partition coefficient used in extraction.

Three classes of adsorbents are common: carbons, inorganics, and synthetic
polymers. The carbons have nonpolar surfaces that adsorb nonpolar solutes. They
are manufactured from a variety of sources, including coke, wood, and coconut
shells. Carbons made from a mixture of sawdust and pumice are often used to remove color from fine chemical solutions. Inorganic adsorbents center on activated
alumina and silica gel, both of which are used as dessicants. These materials have
polar surfaces, and so tend to be more effective for polar solutes. Synthetic polymers include ion exchangers. Although ion exchangers are most often designed to
capture multivalent ions in exchange for monovalent ones, they are often remarkably effective for selectively adsorbing high value-added solutes such as drugs and
pigments. Often, the desorption to regenerate the ion exchanger can be more selective than the original adsorption. Discovering these features normally requires
experiments.

Once we have chosen an adsorbent and measured its isotherm, we will normally
check its effectiveness in a batch adsorption. We will simply drop some adsorbent
into our feed solution, and see how much solute is adsorbed. In analyzing this
batch adsorption, we begin with a mass balance.

$$V y_{10} = V y_1 + W q_1, \tag{6.2-14}$$

where V is the volume of liquid solution, y_{10} and y_1 are the initial and final solute concentrations in the solution, W is the volume of adsorbent, and q_1 is the solute volume per volume of adsorbent. We then consider how the concentrations are related when they are in equilibrium. One such relation is a Freundlich isotherm,

$$q_1 = Ky_1^n, \tag{6.2--15}$$

where K is an equilibrium constant and the exponent n is less than one for a favorable isotherm.

We can solve for the batch concentrations y_1 and q_1 either numerically or graphically, though rarely analytically. But in practice such batch adsorptions are uncommon. One case in which batch adsorption is used concerns the recovery of extracellular products such as antibiotics from a fermentation broth. By dropping the adsorbent directly into the broth, we can adsorb the product directly and avoid the sometimes difficult filtration of the broth. Still, this case is the exception, not the rule. There is often a better way.

This better way to do adsorptions is to put the adsorbent in a packed bed, and to pour the feed solution through the bed. This is the way we will normally perform adsorptions, because it ensures that the adsorbent is in equilibrium with the feed concentration and not with the smaller depleted batch concentration. This means that the expensive adsorbent adsorbs more. The only time not to use a packed bed is if the feed can plug the bed, as in the case of the fermentation broth mentioned in the previous paragraph.

If we feed the product solution into a packed bed, we normally expect the product will adsorb to saturate the adsorbent. The result will be a zone of the bed, saturated with solute, which grows with time. If there were no dispersion in a perfectly homogeneous bed, then the concentration profile would be a step function, moving slowly through the bed. When the bed was totally saturated, the exiting concentration would jump from zero to the feed concentration. Such a jump is called a "breakthrough curve," and it is shown schematically at the top of Figure 6.2–4.

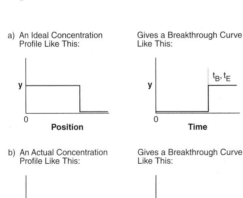

Figure 6.2–4. Breakthrough Curves. The concentration profile within the bed, suggested on the left, leads to the concentrations eluted that are shown on the right. The idealized, sudden step in a) is the limit of the common, gentle rise in b).

Unfortunately, packed beds do not show these ideal step changes in their output. Instead, the concentration profile within the bed is dispersed, as shown at the bottom of Figure 6.2–4. This dispersion can have many causes, including inhomogeneities in packing, Taylor dispersion, and axial diffusion.

Because the breakthrough is not a step function, we will need to use more adsorbent than the minimum needed. One approximate way to estimate this amount is as a "length of unused bed l'," which can be estimated as

$$\frac{l'}{l} = (t_E - t_B)/2t_B, \qquad (6.2\text{--}16)$$

where l is the actual length of the packed bed; the breakthrough time, t_B, is that when the solute starts flowing out of the bed; and the exhaustion time, t_E, is the time when the bed is completely saturated so that the concentration coming out of the bed is the feed concentration. These times are illustrated at the bottom of Figure 6.2–4. Thus building a bed is simple: make an experiment to get the breakthrough curve; find t_B and t_E; and calculate the length of unused bed l'.

We are always amazed that this procedure works, and if you are a novice, you should be too. After all, look at the dispersed concentration profile at the bottom of Figure 6.2–4. Intuitively, we would expect that as the bed gets longer and longer, the concentration profile will spread more and more. Then l' will be bigger and bigger.

Amazingly, this does not happen: l' stays about constant as the bed gets longer. The reason is that the concentration profile within the bed tends to develop and then, holding its shape, moves on through the bed. It does so because of the nature of most isotherms. Most isotherms are favorable; they adsorb more strongly in dilute solution than in concentrated solution. Thus any solute that strays ahead of the profile is adsorbed and thus retarded, and any solute that lags behind tends to flow ahead more quickly. The result is a concentration profile that is self-sharpening. It tends to become more like a step function, and hence makes less severe any dispersion. Thus adsorption tends to correct its own dispersion, and hence becomes one of the key separation processes for specialty products.

Crystallization

The third important separation process for specialty chemicals is crystallization, and its bastard cousin, precipitation. Both these separations involve making solids form in solution. Precipitation is usually a poorly controlled process, done quickly to concentrate the product, to facilitate its isolation. Crystallization is done much more slowly, and aims at dramatic purification. It is often the penultimate step in specialty separation, followed only by drying. Both are reviewed below.

In most cases, we will trigger precipitation just by adding a nonsolvent to the solution produced by the chemical reactions. The nonsolvent is miscible in the solution, but causes solutes – including the product – to precipitate. Such a precipitation, often called a "knock out," occurs because the free energy of the product in solution is increased above that of the solid product.

Nonsolvents normally have a very different polarity to that of the product. For example, if the feed is aqueous, the nonsolvent may be acetone or t-butanol. If the feed has a solvent such as ethanol, the nonsolvent is usually water. If the feed is potentially ionic, the precipitation can be effected by excess salt. Ammonium sulfate and potassium citrate are two common choices as salts. Beyond these generalities, we suggest four heuristics:

1. Precipitation increases as temperature decreases.
2. Precipitation of high molecular weight products is easier than low molecular weight ones.
3. Precipitation tends to be easier if many solutes are present.
4. Precipitation from water is easier when the ionic strength is around 0.1 M.

For more exact results, we must depend on experiment.

Crystallization is a more ambitious process than precipitation, because crystallization tries to purify the product, not just to concentrate it. A crystallization also will aim at large crystals, perhaps around 300 μm. Large crystals are much easier to wash and filter, which are normally the next steps in the separation. Crystallization is one of the most important separation processes for specialty chemicals.

Crystallization depends on three key factors. First, it depends on solubility variation with temperature and solvent composition, an equilibrium factor parallel to the partition coefficient for extraction and the isotherm for adsorption. Second, crystallization depends on the crystal growth rate. Third, crystallization depends on the "cooling curve." Each of these three key factors merits more explanation.

Solubility changes with temperature and solvent composition. Usually, the solubility increases as temperature increases. By reducing the temperature or changing the solvent concentration, we can potentially initiate crystal formation.

The difficulty with this simple picture is that solutions can often contain more solute than that present at saturation. Such supersaturated solutions are thermodynamically unstable. However, they can be metastable; that is, they can remain unaltered indefinitely. This metastability, the result of the surface energy of small crystals, was detailed for microstructured products in Section 5.4. There it was a benefit; now it inhibits purification.

To overcome this thermodynamic barrier, we will often add seed crystals to start the crystals growing in the supersaturated solution. Ideally, these seeds will be of pure product; their number will equal the number of large crystals eventually grown. In fact, in industrial crystallizers, the seeds dominate crystal formation only at the start of the process. At later times, secondary nucleation may become important, especially in larger crystallizers. We will return to this point in the next section. For now, remember that nucleation and supersaturation complicate the effect of equilibrium solubility, still the basic requirement for this separation.

In addition to the first key factor of solubility, crystallization depends on a second key factor, crystal growth rate. Crystal growth is, in many instances, controlled by diffusion, and it is described by the equation

$$dM/dt = kA(c - c^*),$$

(6.2–17)

where M is the crystal mass, A is the total crystal area, k is a mass transfer coefficient, and c and c^* are the solute concentration actually in the solution and at saturation, respectively. Other mechanisms of crystal growth also occur, but the focus on the diffusion-controlled case illustrates the ideas involved.

This rate equation is complicated because the crystal area, A, varies with the crystal mass, M. For simplicity, we will assume spherical crystals. For such a spherical crystal,

$$M = 4r^3\rho/3, \tag{6.2–18}$$
$$A = 4\pi r^2, \tag{6.2–19}$$

and

$$dr/dt = (k_D/\rho)(c - c^*) = G, \tag{6.2–20}$$

where G is the growth rate of a single crystal. Note that this rate is independent of crystal size and linearly dependent on the degree of supersaturation.

We now have two of the three key factors for crystallization. From experiment, we will establish the saturation concentration, the first factor. Also from experiment, we will determine the growth rate, the second key factor. This growth rate should give us crystals large enough to filter.

We now turn to the third factor, the cooling curve. Implicitly, we are using a batch crystallizer containing a supersaturated solution of our product. This crystallizer is normally nothing more than a stirred tank. We will start the process by adding seed crystals. We want to know how fast we should cool the tank to get product crystals of the desired size and purity. Alternatively, if we are crystallizing by adding a nonsolvent, we want to know how fast we should add the nonsolvent. For simplicity, we will discuss only the cooling case in the paragraphs that follow.

In general, we have the greatest chance of a successful crystallization if we aim at a constant crystal growth rate, G. From the analysis above, we see that G is the product of the mass transfer coefficient, k_D, and the degree of product supersaturation, $(c - c^*)$. The coefficient k_D does not change much with temperature. Thus for a constant crystal growth rate, we want to cool so that $(c - c^*)$ is constant.

We can now calculate the cooling curve for constant crystal growth rate. Although the details of this estimate are beyond the scope of this book, the result is worth discussing:

$$(T - T_0)/(T_F - T_0) = (3M_s/M_F)\tau[1 + \tau + (\tau^2/3)], \tag{6.2–21}$$

where T is the time dependent temperature that we seek; T_0 and T_F are the initial and final temperatures, respectively; M_S is the mass of seeds; M_F is the maximum possible crystal mass minus the seeds' mass; and τ is a dimensionless time. This dimensionless time is given by

$$\tau = [(R_F/R_S) - 1]t/t_F, \tag{6.2–22}$$

where R_F and R_S are the final and seed sizes of the crystals, respectively; and t and t_F are the actual time and the final crystallization times. At the start, when time

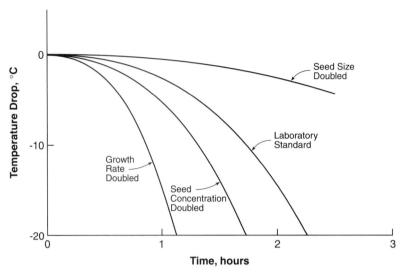

Figure 6.2–5. A Typical Cooling Curve for a Batch Crystallization. To crystallize this steroid, we begin by cooling slowly and then cool faster and faster. We would get roughly similar results if we crystallized by adding a nonsolvent.

is small, we should cool slowly, linearly. As time gets larger, we should cool faster and faster to get large pure crystals. An example of such a curve for a steroid is given in Figure 6.2–5. A similar strategy can be used when we are not cooling, but precipitating by slowly adding a nonsolvent.

This long section had two goals. First, we wanted to explore the sequence in which any separation processes should take place. We argued that product concentration should normally take precedence over purification. Second, we wanted to consider which separations are likely to be important for specialty chemicals, and to quickly review the three most important. We have now finished this review of extraction, adsorption, and crystallization. To illustrate these ideas, we now turn to an example.

EXAMPLE 6.2–1   PENICILLIN PURIFICATION

This classic process is the model for a huge group of antibiotics, including cephalosporins, which are based on β-lactams. For the penicillins, the basic structure is

These molecules can be made either chemically or microbiologically. In the microbiological route, mutants of *Penicillium chrysogenum* are grown in 100,000-L aerated fermenters that are charged primarily with lactose, corn steep liquor, and

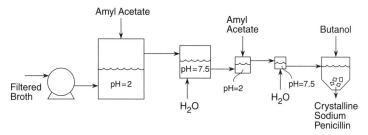

Figure 6.2–6. A Schematic Process for Penicillin Purification. Aqueous pencillin at pH 2 is extracted into amyl acetate and then backextracted into water at pH 7.5. After this process is repeated, the product is crystallized from a water-butanol mixture.

calcium carbonate. After about 7 days, the broth contains perhaps 80 mg of penicillin per liter of broth.

Suggest how the penicillin can be isolated and purified.

SOLUTION

The key to this purification is the recognition that these materials are carboxylic acids. When the pH is above about 5.5, the COOH group ionizes to COO⁻ and the penicillin becomes water soluble. When the pH is below 5, the COOH group remains protonated, and the penicillin is more soluble in organic extraction solvents.

Hence, the purifications used for penicillin rely heavily on extraction. A typical process is shown in Figure 6.2–6. This first step is to separate the penicillin containing broth from the large biomass of micro-organisms. Because normal filters tend to plug, this separation often involves adsorption of the microbes on diamataceous earth (Filter-Aid) and then filtration. The clarified broth is acidified and then extracted with amyl acetate. Because the acid form of the penicillin is less stable, this extraction should be as fast as possible. Then the amyl acetate is extracted with water at pH 7.5, so the product moves back into the water. This entire process is repeated until the penicillin is concentrated perhaps 100 times. Finally, butanol is added to the aqueous penicillin solution to precipitate crystals of sodium or potassium penicillin.

This process is a simplification of what is actually done. In fact, the first amyl acetate extract is decolorized by adsorption on activated carbon. The last aqueous extract may be dried as a crude product before it is redissolved and crystallized. Still, this purification is an excellent example because it depends on recognizing one key chemical fact: penicillin is a carboxylic acid.

6.3   Specialty Scale-Up

By this point, we know what we are going to make, how we are going to make it, and how we will purify this specialty product. We now need to make commercial quantities of a purity equal to or better than our small-scale studies.

We have to scale-up our process. This scale-up is different from that used for commodity materials. There, we must design a truly dedicated process. We will be eager to invent new, more effective forms of chemical reactors. We will be happy to consider new separation processes. We will judge our success purely on chemical grounds: is what we produce of the same or higher purity to that which we made before?

For specialty materials, our challenge is different in two important ways. First, we will probably already have the equipment that was designed, not for our product, but for a spectrum of past products. We must make our new product in this generic equipment. We do not want to build a new style of reactor; we will almost certainly just use our old, familiar stirred tanks. We do not want to try separation methods beyond those we already have. For example, we may be happy to buy a new adsorbent to use in our existing adsorption columns; but we probably will not want to try a nonstandard separation, such as sublimation or zone melting.

The second important difference for these products is specific to pharmaceuticals. In many cases, the material used in clinical trials will have been made in the laboratory scale process. This is the material, and the process, that which will have been approved. If we were to decide on another type of reactor or another reaction solvent, we may need to make a new clinical trial and request a new approval. Because this is slow and expensive, we will almost certainly want to avoid it. We are going to stay much closer to our laboratory procedure than we would ever do for a commodity chemical.

In scale-up of a specialty product, we will want to keep process variables in the same proportions as they were in the laboratory. We want the same rate constant in the full-scale reactor as in our laboratory reaction. We want the same heat transfer in production as in the laboratory. We want the same purification at the large scale as at the small scale.

Getting the same performance on the larger scale as the small scale is the subject of this section. In discussing this subject, we parallel the earlier topics in the chapter. We start with chemical reactions, including those controlled by chemical kinetics and by mass transfer. We then explore the three key separations of extraction, adsorption, and crystallization. In all of this discussion, we focus on batch processes in generic equipment, for these are the norm for specialty chemicals.

REACTOR SCALE-UP

The scale-up of chemical reactors involves three topics: chemical kinetic control, mass transfer control, and heat transfer. When reactors are controlled by chemical kinetics, the scale-up is simplest. For a first order reaction, the concentration, c_1, is given by

$$c_1/c_{10} = e^{-kt}, \tag{6.3--1}$$

where c_{10} is the initial concentration, t is the time, and k is the first order rate constant, with dimensions of reciprocal time. If the reaction is isothermal, nothing in this equation depends on scale. A large reactor 100 times bigger than the laboratory unit will behave just like 100 laboratory units run in parallel. If we got 99% conversion in the laboratory in 80 minutes, then we will get 99% conversion in the plant in 80 minutes.

Chemical engineers sometimes dignify this conclusion by rephrasing it in terms of dimensionless groups. These groups are useful for intimidating chemists, who feel that they should know what the groups mean, but do not. Groups with Germanic names – Schmidt, Prandlt, Thiele, Damkohler – are especially good intimidators because they echo vanished heroes of German chemical science. However, in batch reactions, dimensionless groups have only marginal value. For example, in this case, we could assert

$$c_1/c_{10} = e^{-\tau}, \tag{6.3-2}$$

where τ is a dimensionless time equal to kt. For 99% conversion, we want the dimensionless concentration (c_1/c_{10}) to equal 0.01, so the dimensionless time τ must be 4.606. We do not feel this puffery helps much.

With or without dimensionless groups, reactor scale-up is rarely as simple as for the first order, isothermal, chemical kinetically controlled limit. In particular, for the case of a mass transfer controlled reaction,

$$c_1/c_{10} = e^{-k_D a t}, \tag{6.3-3}$$

where k_D is the mass transfer coefficient and a is a reactant surface area per volume. Just as before, we want to scale up the reaction, keeping $k_D a$ constant; unlike before, $k_D a$ does depend on the size and operation of the reactor. This is true whether the reaction is isothermal or adiabatic. Anyone who has tried making 5 kg of tiramisu in one batch will understand how dependant on scale mass transfer can be.

The exact dependence of $k_D a$ on scale is a function of the specific situation involved. This dependence is best illustrated by two examples. The first example is aeration of a stirred tank, where from experiments we know that

$$k_D a = f(P/V, v_g), \tag{6.3-4}$$

where (P/V) is the stirrer power per volume and v_g is the superficial air velocity, that is, the volume of air used per cross sectional area of the tank. We will scale up the reaction by using the same power per volume and the same superficial air velocity. We will also keep the relative dimensions of the tank the same: the ratios of the tank height to diameter, and of the diameter to the stirrer's impeller length should be the same in the production reactor as in the laboratory reactor. If this is done, $k_D a$ should be the same in both cases.

The second example involves slowly adding a limiting reagent to a chilled solution of excess reagents. Again, we believe the reaction is not governed by chemical kinetics, but only by the rate of mixing. It is like the mass transfer controlled case of aeration just discussed, but here there is no interface between gas and liquid. Still, the process fits first-order kinetics, given by Equation 6.3–3.

As before, we want to transfer the chemistry from a laboratory reactor into a production-sized batch reactor. The key is again the rate constant k, which must be inversely related to the time for mixing. This mixing must involve two scales: the large-scale mixing that is caused by the impeller, and the small-scale micromixing that is due to diffusion. Although we will stir hard to make the large-scale mixing as fast as possible, we cannot do much about speeding up diffusion. We do know that the micromixing time will be

$$[\text{micromixing time}] \propto 1/k_D a \propto l^2/D, \tag{6.3–5}$$

where l is the eddy size caused by the large-scale mixing. Fluid mechanical arguments suggest that

$$l \propto \left(\frac{\rho v^3}{P/V}\right)^{1/4}, \tag{6.3–6}$$

where ρ is the liquid density, v is the kinematic viscosity, and (P/V) is the stirring power per volume. Thus

$$k_D a \propto D\left(\frac{P/V}{\rho v^3}\right)^{1/2}. \tag{6.3–7}$$

We are stuck with the values of D, ρ, and v, for they are physical properties of our system. Thus as before, we want to run at the same power per volume at both small and large scales.

We now turn to the third issue in reactor scale-up, that of heat transfer. Both the two earlier scale-up cases of chemical kinetic control and of mass transfer control have assumed the reactions are isothermal. In fact, although they may be run close to isothermally in the laboratory, they frequently will involve considerable evolution of heat. This is because we tend to use high energy reactants to get selective chemical reactions. We can justify these expensive reactants because we are making high value-added products. This means that we must get rid of substantial heats of reaction.

Doing so is more difficult at large scale than in the laboratory. This is because the amount of heat removed depends on the reactor's surface area, and the amount of heat produced depends on the reactor volume. The ratio of area to volume gets smaller as the reactor gets bigger.

In most cases, we cannot deal with this problem without changing the way in which we transfer heat. By making an energy balance on an instantaneous adiabatic reaction in our reactor, we could find that the temperature jumped from an initial ambient T_0, to a new hotter T_R. The reactor would then cool according to the relation

$$\frac{T - T_0}{T_R - T_0} = \exp\left(-\frac{Uat}{\rho \hat{C}_p}\right), \tag{6.3–8}$$

where U is the overall heat transfer coefficient, ρ is the solution's density, and $\hat{C}_p$ is the specific heat capacity. (Note $\rho \hat{C}_p/Ua$ is the characteristic time for cooling the reactor.) Although we cannot change ρ and $\hat{C}_p$ from the smaller reactor to the larger one, we may be able to increase U in the larger reactor. This occasionally works. More often, however, we will be forced to take refuge in another mechanism of heat transfer in order to keep the big reactor cool. One common answer is to reduce the pressure in the reactor, so that the solvent can boil.

In addition to these concerns of reactor scale-up, we will also have problems of scaling up our separation processes. We turn to these next.

SEPARATION SCALE-UP

Just as we needed to scale up the laboratory reactor, we will need to scale up the separation processes. Just as we used generic reactors, not optimized for our specific product, so we must use generic separation equipment. This equipment will probably be designed for the largest amount of product we have made in the past and not for our new product.

We will restrict our discussion to the scale-up of extraction, adsorption, and crystallization, commonly the three most critical separation processes for specialty chemicals. We give our conclusions before we begin. The scale-up of extraction is straightforward. The scale-up of adsorption is tricky and normally requires some compromise. The scale-up of crystallization invites prayers to St. Jude, the patron saint of lost causes.

To see why scaling up extraction is straightforward, we return to the analysis of staged extraction in the previous section. This analysis depends only on the variable

$$\frac{mL}{H} = \frac{(\text{partition coefficient})(\text{amount of solvent})}{(\text{amount of feed})}. \tag{6.3–9}$$

Because the partition coefficient does not change with scale, this says that we just need to keep L/H constant. If we have 100 times more feed in production than in the laboratory, we will need 100 times more solvent. The only source of potential trouble in this sanguine picture is the time necessary to reach equilibrium after the mixer. This implies mixing in the same way, that is, keeping the stirrer power per mixer volume the same. In practice, this is rarely a problem. No wonder process engineers love extraction.

Scaling up adsorption is more difficult. The difficulty arises because we seek an efficient bed, with a sharp breakthrough curve. This is most easily achieved by using a bed of small particles, which in turn implies a large pressure drop across the bed. In the extreme case of analytical chromatography, the bed's particles may be around 10 μm in diameter and the pressure drop 20,000 kPa. Running large columns with this high a pressure drop is normally impractical.

A large-scale adsorption bed must therefore use much larger particles. A typical size chosen is 300 μm. We will normally need to make experiments with these larger particles, understanding that their use implies a much less abrupt breakthrough curve. This in turn implies a larger length of unused bed, and so a less efficient use of our adsorbent than in the analytical columns.

At this point, present industrial practice seems to split into a conservative strategy and a more ambitious, risky one. In the conservative strategy, new laboratory experiments with the large particles are used to define the bed length and the feed velocity that will be used in the large-scale separation. If the feed has a viscosity μ close to that of water, the velocity v in the bed will be given by

$$v = \frac{\Delta p d^2}{150\,\mu l}\left[\frac{\varepsilon^3}{(1-\varepsilon)^2}\right], \tag{6.3--10}$$

where Δp is the pressure drop, d is the particle diameter, μ is the feed viscosity, l is the bed length, and ε is the void fraction. We may decide on a feed velocity of 10 cm/min in a bed of 100 μm particles which is 1 m long, and has a void fraction of 0.3. In this case,

$$\frac{0.1 \text{ m}}{60 \text{ sec}} = \frac{\Delta p(100\times 10^{-6}\text{ m})^2}{150(10^{-3}\text{ kg/m sec})1\text{ m}}\left[\frac{0.3^3}{0.7^2}\right].$$
$$\Delta p = 500 \text{ kPa}. \tag{6.3--11}$$

Higher pressure drops can deform the packing and hence dramatically reduce the flow. These data would be obtained at small scale, using a bed of perhaps 1 cm in diameter.

If we follow the conservative strategy, we then increase the scale of the adsorption merely by increasing the diameter of the column. We use the same sized particles, in the same length of bed. We use the same feed velocity. We just make the bed fatter. For example, if we wanted to scale up the adsorption described in the previous paragraph by 1000 times, a characteristic value, we just go from a 1-cm bed to a 32-cm bed, thus increasing the bed's cross sectional area and hence the total flow about 1000 times.

The conservative strategy can work well, but it can also produce some bizarre looking adsorption columns. In extreme cases, these columns have had a diameter over twenty times their length. Such "pancake" columns may require complex feed distributors and elaborate desorption procedures. Such complexity may not be easily achieved in the generic equipment already available in our pilot plant.

In these cases, we may prefer the more ambitious, risky strategy. In this strategy, we will increase the bed length and the fluid velocity when we go to larger scale. We will therefore be forced to increase the particle size in order to keep the pressure drop modest. We will accept the less abrupt breakthrough curve such a strategy for scale-up implies.

The difficulty with this risky strategy is that the less abrupt breakthrough may be difficult to predict, for it depends on the mechanism of the steps controlling the adsorption. A detailed discussion of these complex steps is beyond the scope of this book. The prediction of these curves has improved dramatically in the last decade. Nonetheless, we recommend that no scale-up be made without experiment. Normally we should expect that the breakthrough will be much less abrupt.

So far, we have shown why extraction scale-up is routine, and how adsorption scale-up can be complicated. We now turn to the scale-up of the third key separation process, crystallization.

Crystallization is the most difficult process to scale up successfully. At first, we may have trouble seeing why. If we effect crystallization by cooling, then the key equation in our analysis is Equation 6.2–21. When we first look at this equation, the only variable that depends on scale is the final mass of the crystals, M_F. For a successful scale-up, we want to keep (M_s/M_F) the same. Thus if we are scaling up 1000 times, we will want to use 1000 times more seeds. We will want to vary the temperature versus time as we did in the laboratory case.

In passing, we might ask where the product's chemistry enters this equation. After all, the equation seems to give temperature versus time, without reference to particular chemistry. In fact, the product chemistry is concealed in the total crystal mass M_F, given by

$$M_F = (T_0 - T_F)V\{dc(\text{sat})/dt\}, \tag{6.3–12}$$

where T_0 and T_F are again the initial and final temperatures; V is the crystallizer volume; and the quantity in braces is the change in solubility with temperature, which we take as constant. Note that this solubility change is where the product chemistry enters our analysis.

Returning to the problem of scale-up, our basic strategy for a large-scale crystallization is to cool so that the temperature varies with time just as in our successful laboratory crystallization. Such a strategy is a good first step for scaling up the crystallization. However, this step is often not completely successful for two reasons. First, we may not be able to cool the large-scale process in exactly the same way as we cooled the small-scale process. Second, we may have different nucleation mechanisms influencing the large-scale process than those that dominate the small-scale one.

The large-scale cooling can differ from the small-scale cooling primarily because the area per volume for heat transfer changes. To see how this effect of scale enters, we first note that we are assuming that the temperature is the same throughout the crystallizer. In general, it will not be. We will generally cool the crystallizer contents by cooling the crystallizer walls, so the walls must be cooler than the bulk. Near these cool walls, supersaturation, crystal growth, and especially nucleation will all be greater than in the bulk of the crystallizer.

This inhomogeneity will be more severe for a large crystallizer than for a small one. To see why, we can make an energy balance on the crystallizer:

[energy change in crystallizer contents] = [energy removed by cooling],

$$\frac{d}{dt}[\rho \hat{C}_p VT] = Ah(T - T_{\text{wall}}), \tag{6.3–13}$$

where T and T_{wall} are the temperatures of the crystallizer's contents and walls, respectively; ρ and $\hat{C}_p$ are the content's density and specific heat capacity; V is the volume of the contents; A is the crystallizer area in contact with the product solution; and h is the individual heat transfer coefficient between the warm contents and the cool wall. We can rearrange this equation as

$$(\rho \hat{C}_p/h)[dT/dt] = (A/V)(T - T_{\text{wall}}). \tag{6.3–14}$$

Everything on the left-hand side of this equation is fixed. The density ρ and the heat capacity $\hat{C}_p$ are just physical properties, and the heat transfer coefficient h does not change much. The derivative dT/dt does change with time, but its value is dictated by the cooling curve found from the small scale, laboratory experiments.

Thus we want to keep the right-hand side of this equation fixed as well. We cannot do it. As the batch gets bigger, the area per volume (A/V) gets smaller. To reach the value dictated by the left-hand side of this equation, we are thus forced to increase ($T - T_{\text{wall}}$). When we do so, the inhomogeneities near the wall become more severe. These more severe inhomogeneities are one reason that the scale-up of crystallization is risky.

The second reason that crystallization is difficult to scale up centers on nucleation, but of a different sort. So far, we have implicitly assumed that nucleation occurs as the result of seeding and that the number of crystals that eventually form equals the number of seeds that we have added. Thus we have implicitly assumed that the only nucleation is heterogeneous, at the time when we add the seeds.

In fact, nucleation can occur by two other important mechanisms: homogeneous and secondary. Homogeneous nucleation occurs without seeds, as the result only of supersaturation. It requires larger supersaturation than is normally present in the bulk of the solution inside the crystallizer. Such larger supersaturation may occasionally occur near the colder walls of the larger crystallizer. Still, homogeneous nucleation is rarely an important mechanism at any scale of industrial crystallization.

In contrast, secondary nucleation is important industrially, especially at larger scale. Secondary nucleation can occur as a result of high shear, from crystal collisions, or as a result of crystal formation on the walls. The high shear of a supersaturated solution flowing past a growing crystal may dislodge crystal embryos or nuclei that would otherwise be incorporated into the growing crystal. These new embryos then grow into new, separate crystals. This mechanism is more important at large scale because bigger crystallizers require locally higher shear for the same mean power input.

Alternatively, secondary nucleation can result from crystal collisions. In the extreme case, these collisions may cleave a large crystal into smaller fragments, each of which grows into a larger crystal. Collisions with the impeller also appear to generate nuclei, sometimes in surprisingly large numbers. This may result from the removal of small dendrites from the crystal surface. Again, this is more important at larger scale because local shear is higher. Finally, nucleation and growth on the crystallizer walls is to be expected: rough spots on the wall supply nuclei, and the colder wall means potentially a higher supersaturation. At larger scale, the walls are often colder, as discussed above.

Our first reaction to this concern with increased secondary nucleation may be to shrug our shoulders. Why should we care? Our goal is to make crystals, and secondary nucleation just makes more crystals. To be sure, the increased number of crystals means that the average size is smaller, but smaller crystals tend to be purer because they are less likely to contain defects.

In fact, we should be concerned if the large-scale secondary nucleation produces large numbers of smaller crystals. Although these crystals may be somewhat purer, they will be much harder to filter out of the mother liquor. Just think: flow through a packed bed of small crystals will be laminar, and hence proportional to the square of particle size. If the particles are five times smaller, the time to filter a given volume will be 25 times longer. This can cripple a large-scale crystallization. This is why the scale-up of crystallization can so frequently produce unpleasant surprises.

So far, we have discussed only crystallization caused by cooling a product solution, and not by adding a nonsolvent to this product solution. Doing so involves an analog to the "cooling curve" where the supersaturation is now governed by solvent addition, not by cooling. At large scale, the driver will not be the cold crystallizer wall, but rather the slow mixing of the nonsolvent. We are not familiar with careful studies of supersaturation caused by large-scale mixing of a nonsolvent. We suspect that many such studies exist, but are hidden in unpublished technical reports of pharmaceutical companies.

EXAMPLE 6.3–1   REACTING SUSPENDED STEROIDS

You are reacting a suspension of steroid particles, about 2.60 μm in diameter, with butyl lithium in tetrahydrofuran at $-20°$C. You believe that the reaction is mass transfer controlled, with the mass transfer coefficient k_D given by the correlation (Boon-Jong et al., 1978)

$$k_D d / D = 0.46 (dd'\omega/\upsilon)^{0.28} (gd^3/\upsilon^2)^{0.17} (M_T/\rho d^3)^{-0.011} (d'/d)^{0.02} (\upsilon/D)^{0.46},$$

where d is the particle diameter; D is the reagent diffusion coefficient in the liquid; d' and ω are impeller diameter and speed, respectively; υ and ρ are the kinematic viscosity and density of the liquid, respectively; g is the acceleration that is due to gravity; and M_T is the particle mass.

You want to scale up the reaction 1000 times using the same size particles. How should you proceed?

SOLUTION

This is a good example of a problem with extraneous information. If you look at the correlation, you see that the only variables you can control are d' and ω. But

$$k_D \propto (d'\omega)^{0.28} (d')^{0.02}.$$

If we are scaling up 1000 times with a geometrically similar reactor, then d' increases ten times. Thus to keep k_D the same, we should decrease ω about eight times. This is close to scaling at constant Reynolds number $(dd'\omega/\upsilon)$.

EXAMPLE 6.3–2   SCALING UP A LINCOMYCIN ADSORPTION

We are adsorbing both lincomycin A and B from a clarified fermentation beer onto a modified dextran resin. The resin, which can stand pressure drops up to 1000 kPa, shows a highly favorable isotherm for these products. In the laboratory, we have run the beads in a 1.6-cm-diameter tube packed to a depth of 34 cm. With a pressure drop of only 60 kPa, we get a lincomycin breakthrough at 46 min, and an exhausted bed at 62 min.

We want to run this system in an existing pilot plant adsorption bed that is 30 cm in diameter. We have already operated this bed by using a pressure drop of 410 kPa. How much can we scale up this process? How should we operate to scale up a factor of 5000 times?

SOLUTION

To begin, imagine that we operate the 30-cm-diameter bed under exactly the same conditions as the laboratory bed. Because the pressure drop and the bed length are unchanged, the velocity, the breakthrough time, and the exhaustion time are all the same. Thus the gain in capacity is solely due to the gain in the relative cross sectional areas:

$$[\text{gain in scale up}] = \left[\frac{\text{cross section of pilot bed}}{\text{cross section of laboratory bed}}\right]$$

$$\frac{\pi/4(30\text{ cm})^2}{\pi/4(1.6\text{ cm})^2} = 350.$$

This reasonable increase is smaller than we seek.

As a second alternative, we can operate at the same velocity but in a deeper bed. In the laboratory, we got satisfactory results in a bed 34 cm deep, using a pressure drop of 60 kPa. Because we know that we can increase the pressure drop to 410 kPa, we can use a deeper bed:

$$[\text{pilot bed depth}] = [\text{laboratory bed depth of 34 cm}]\frac{410\text{ kPa}}{60\text{ kPa}} = 230\text{ cm}.$$

Because the isotherm is favorable, the length of unused bed is constant. For the laboratory bed, the length of unused bed l' is just

$$l' = l\left(\frac{t_E - t_B}{2t_B}\right),$$

$$= 34\text{ cm}\left[\frac{62\text{ min} - 46\text{ min}}{2(46\text{ min})}\right] = 6\text{ cm}.$$

Thus the increased capacity is now

$$\text{gain in scale up} = \frac{\text{volume used in pilot bed}}{\text{volume used in laboratory bed}},$$

$$= \frac{\pi/4(30\text{ cm})^2(230\text{ cm} - 6\text{ cm})}{\pi/4(1.6\text{ cm})^2(34\text{ cm} - 6\text{ cm})},$$

$$= 2800.$$

This is a substantial increase, but still less than the factor of 5000 we had hoped for.

To go still higher, we will need to take risks tempered by additional experiments. If possible, we still want to use the same bed velocity. Because the adsorbent is said to stand a pressure drop of 1000 kPa, we could investigate running at the same velocity in a still deeper bed. To do so, we must make sure that our pilot column can also stand the higher pressure. Alternatively, we can increase the adsorbent diameter and hence the velocity through the bed. Doing so will almost certainly increase the length of unused bed, perhaps dramatically. Before we make this more radical change, we will need more laboratory experiments using bigger velocities past bigger particles.

6.4 Conclusions

In many ways, this chapter will seem the most familiar of this book for chemists and chemical engineers. It begins with a specialty product selected for manufacture. The chemistry of this specialty product is known. Often, it will be chemically well defined, with a known molecular structure. Typically, it will have a molecular weight between 300 and 1000 daltons. It will be nonvolatile, but soluble in a variety of solvents. Often, it will have some aqueous solubility, which will frequently be a function of pH. In many cases, the product may be somewhat unstable, especially as the temperature is increased.

In this chapter, our target has been to make commercial quantities of this product. We use both reaction engineering and separation processes in planning production. As part of the reaction engineering, we will normally make measurements that supply a reaction rate constant. These reaction rate constants will normally be first order, though they may occasionally be zero order. The first order rate constants for making our target product will be the result either of chemical kinetics or mass transfer. Although we are not sure, we believe that mass transfer kinetics are more common for specialty products than for commodity chemicals. We suspect that this is because the higher added value of specialties encourages use of more expensive and more active reagents.

The separation processes chosen for specialty product manufacture will commonly include extraction, adsorption, and crystallization. Other unit operations will normally be involved, especially filtration and drying, but these will be relatively easily accomplished. Extraction will often be used less to purify the product than to concentrate it. Adsorption tends to be the workhorse for purification. Crystallization often serves as a final purification or "polishing," and it is particularly important for pharmaceutical products.

This brief litany misses a key point of specialty chemical manufacture. This key point is that we do not plan to make much product. The market is small, though the price per kilo is large.

Although the high added value makes us eager to manufacture our product, the small market size will make us unwilling to commit dedicated equipment to this task. We will use the same equipment for this product that we use for other products. We will make the chemical synthesis in small batches, often working in an 8-hour day. We will not be running three shifts, around the clock. We will make a batch of product, which will supply the market for a few months or a year. We will store the product and sell out of this inventory. When the inventory gets low, we will make another batch.

In many ways, then, specialty chemical manufacture is like gourmet cooking. We have a few pots and pans, and we will make expensive products in near personal batches, as they are needed. We are not in the equivalent of the commercial baking business, continuously producing new loaves. We are not making the bread of life. We are making the specialty chemical products which are the jam of life. The guidelines for making that jam have been the subject of this chapter.

FURTHER READING

Bailey, J. E. and Ollis, D. F. (1986). *Biochemical Engineering Fundamentals*, 2nd ed. McGraw-Hill, New York, ISBN 0070032122.

Belter, P. A., Cussler, E. L., and Hu, W. S. (1988). *Bioseparations*. J. Wiley, New York, ISBN 0471847372.

Blanch, H. W. and Clark, D. S. (1997). *Biochemical Engineering*. Marcel Dekker, New York, ISBN 0824700996.

Boon-Long, S., Laguerie, C., and Couderc, J. P. (1978). Mass Transfer from Suspended Solids to a Liquid in Agitated Vessels. *Chem. Eng. Sci.* **33**, 813–819.

Schmidt, L. D. (1998). *The Engineering of Chemical Reactions*. Oxford University Press, Oxford, ISBN 0195105885.

7

Economic Concerns

By this point, we have largely completed our description of product design. We began by discussing the customers' need for a product, and how this need could be converted into more specific quantitative specifications. We then described generating ideas to fill these needs. Because good ideas can be hard to find, we need a lot of alternatives. Because a lot of alternatives are too many to evaluate completely, we outlined and edited these ideas, using matrix screening methods to decide on our best alternatives. We next made more detailed scientific and engineering calculations of these alternatives, finally reaching our best choice. If the new product is a chemical, we will probably manufacture it in batch, using generic equipment that is not designed for this particular chemical. If it is a device for chemical change, we will use manufacturing techniques established by mechanical engineers.

But we have not talked about money. To be sure, we have repeatedly mentioned that costs and prices are important, but we have not included them in any systematic way. Those with business training may find this omission overwhelming. These business persons may argue that chemical product design must include detailed discussions of financial issues. They may point to the large and detailed business literature, which includes examples aimed both at beginners and at accomplished professionals. They are correct to stress the value of these resources.

At the same time, our goal in this book is to focus on the chemistry and engineering that are central to chemical product design. Making detailed financial projections for a product that implies scientific nonsense is clearly folly. If the product does not work, what good is it? Moreover, in our judgment, the resources, including both people and literature, are much more developed for the financial aspects of product design than for the technical aspects of this same design, particularly in the area of the chemical industry. This is part of the reason why we have written this book.

We do want to include the briefest outline of the finances of chemical product design. We do so partly so that chemists and engineers will understand how the financial arguments are likely to be made. After all, we need the financial people to be our allies, so we need to know at least part of their language.

We also need to discuss the economics of chemical products because it is phrased in different terms to the economics of chemical processes. At present, most training in industrial chemistry describes the production of commodity chemicals. Much less describes specialty chemicals. Many chemical engineering students may have already studied chemical process design before considering chemical product design. Many with industrial experience in petrochemicals may have taken new jobs in areas such as pharmaceuticals.

For all these groups, we want to supply a précis of economic issues. We begin in Section 7.1 with a more detailed description of the differences between products and processes. This distinction centers on how much is made. In Section 7.2, we describe the economics of chemical processes. This branch of economics implies designing and building a chemical plant dedicated to the large-scale production of one product. In Section 7.3, we discuss product economics, especially in terms of net present value. This allows us to quickly estimate how much money we can hope to make from a particular product. This chapter is intended only as the briefest of outlines, but one that illustrates some critical differences between the economics of chemical products (here exemplified by specialty chemicals) and commodity chemical processes. These differences are important in accurately assessing the economic potential of chemical products, as opposed to chemical processes, with which readers of this book are more likely to be familiar.

7.1   Product versus Process Design

The rationale for this book implies that growth in the chemical industry will depend on new products as well as new processes. This implication is certainly consistent with most current corporate strategies. With a few exceptions, the large chemical companies are tending to deemphasize the commodity chemical business to focus on specialty chemicals and other chemical products. In some cases, these companies have left the commodity chemical business altogether. In other cases, companies planning to make commodities have become private, presumably because they feel more confident of withstanding market cycles. In still other cases, especially in continental Europe, public companies plan to continue manufacturing commodities, though usually with an increased commitment to specialty products.

In this situation, we do well to ask what the difference is between specialty and commodity products. We are especially interested in how specialty products and commodity products are judged financially. In this section, we explore these differences. In what follows we concentrate on the economics of specialty chemicals, although throughout the book we have emphasized that specialties are only a subset of chemical products. We do this because specialty chemicals provide the most direct and obvious comparison with commodities. We do believe that the economics of specialties are a good example for other chemical products and so the discussion is relevant to all forms of chemical products.

COMMODITY PRODUCTS

To begin this discussion, we suggest differentiating between specialty and commodity products on the basis of three criteria: how much product is made, what equipment is used, and which producer makes the most money. We begin with commodity chemicals:

1. *How much is made?* Commodities are normally made in quantities greater than 10,000 tons per year.
2. *What equipment is used?* Commodities are normally manufactured in dedicated equipment that is operated continuously.
3. *Which producer makes the most money?* As a general rule, the one with the lowest manufacturing cost will be the most profitable.

These generalizations deserve discussion.

The choice of 10,000 tons per year is the rough consensus of those in the chemical industry. Many of these chemicals are made from petroleum. They provide most of the examples in the chemical engineering undergraduate curriculum. Although some inorganics are mentioned, usually in discussions of stoichiometry, the vast bulk are petrochemicals. Ethylene, butanol, and vinyl chloride are good examples.

These commodity organics are almost always made in very large chemical plants dedicated to making a single product. This has been the case since about 1970. Vinyl chloride is a good example: between 1964 and 1972, the number of U.S. producers of vinyl chloride shrank by 70%, but the median plant size went up by more than a factor of ten. The reason is that the cost of a chemical plant is roughly proportional to the two thirds power of its capacity. Although the reasons for this are complex, we can rationalize it by saying that the plant cost is proportional to the amount of steel needed, which is roughly proportional to the equipment's surface area, which is proportional to the two thirds power of the equipment's volume.

Thus if we want to make commodity chemicals, we must be prepared for a huge capital investment. This is why the capital investment per employee is larger in commodity chemicals than in any other industry. It is also why we are forced to operate continuously. We cannot afford to ever have so much expensive equipment sitting idle. We will normally be most profitable if we operate all day, every day of the year.

Most chemical commodities have been made for decades, using technology that does not change much from one year to the next. All commodities are sold into competitive markets. Moreover, the commodities are chemically well defined. For example, there is no difference whatsoever between propylene made by Exxon or by Hoechst. There is no difference between urea made by W. R. Grace and Cargill. These markets will be highly competitive.

As a result, the very large, dedicated chemical plants must be run efficiently to make a profit. This is why process optimization and computer control have been so important to the commodity chemical business. We are faced with mature technology in a huge but competitive market; we have very limited options for growth. No wonder that small incremental advantages discovered with computers have recently been so important.

SPECIALTY PRODUCTS

We can describe specialty products in terms of the same three criteria used for commodities: how much is made, what equipment is used, and which producer makes the most money. The brief answers for specialties are as follows:

1. *How much is made?* Most specialties are made in quantities less than 10 tons per year.
2. *What equipment is used?* Specialty chemicals tend to be made in generic equipment.
3. *Which producer makes the most money?* The company that first markets the product tends to get 70% of the total sales.

Again, these generalizations merit discussion.

As discussed in detail in Chapter 6, the small amounts made usually imply manufacture in batch processes, not continuous processes. These batch processes often will not run twenty four hours per day, and rarely run all year. Instead, production is usually in "campaigns" in which the product is made for a few weeks, and then stored as inventory. In extreme cases, in the drug industry, the inventory may never exceed a few hundred grams. When the inventory gets small, another campaign is started.

These small amounts are produced in generic equipment, which is used for several different products. In the pharmaceutical industry, we frequently find the equipment used for as many as twenty different products. The generic equipment is often a gaggle of stainless steel reactors, stills, extractors, and, importantly, holding tanks. These are not optimized for any one specific product. Instead, they are designed for flexibility, for many different chemistries.

The generic equipment is often operated in ways that closely imitate the original product discovery experiments in the laboratory. The reactors tend to be large-scale analogs of round-bottom flasks. Reagents are added to the flask, and the reaction is carried out, often in liquid solution. Then other solvents are added, for example, to precipitate the reaction product. The remaining solution is decanted; different solvents are added to redissolve the product; and new reagents are metered in to start a different reaction. Thus in specialty manufacture, different reactions do not use several different reactors, but rather one pot, cooked again and again.

Many engineers react with horror when they first see these specialty syntheses. They immediately see many ways to make process improvements, like getting faster conversions or using less carcinogenic solvents. These changes are rarely welcomed, because these complex reactions may be difficult to control. When the process is working, those in production are always suspicious of any "improvements."

In some cases, sensible improvements may be legally difficult to implement, especially in the drug industry. This is a consequence of reasonable but conservative regulations by the United States' Food and Drug Administration (FDA). Usually, early drug manufacture is rushed to produce enough material for clinical tests. These clinical tests are tedious, the step that tends to limit new drug development. If the tests are successful, the FDA will with deliberate speed approve the drug, but only for the process used to make the original material. The fact that the drug can be made in another way, more efficiently and without carcinogenic solvents, does not matter. The fact that the final drug is chemically well defined, of known molecular structure, and equal or better purity, does not matter. If we change the process, we must again petition the FDA for a new approval.

Finally, we re-emphasize that the producer who makes the most money is usually the first to market, and not necessarily the low cost producer. Specialties tend to be high value-added products, selling for much more than the cost of raw materials. Antibiotics are an excellent example: although they can sell for in excess of 100$/kg, they are commonly made from agricultural waste streams. Specialty products are clearly very different to commodity products.

We should note that commodity chemicals are already strongly emphasized in academic chemical engineering. We teach details of distillation, which is central to commodity petrochemicals, but often tangential to specialties such as pharmaceuticals. We only superficially describe adsorption, a common purification method for pharmaceuticals that is infrequently used for commodity petrochemicals. Most of all, we teach economic tools that imply continuous dedicated plants, and de-emphasize the tools for specialty products. These different economic tools are summarized in the next sections.

7.2   Process Economics

In the type of economics traditionally taught to chemical engineering students, we are typically planning to make a chemically well defined product that is already made by others. Because of this competition, we will know what we can expect as a price for our product. We will be most concerned with the cost of producing the product.

One way in which we can approach the economic challenges of commodity chemicals is to consider the process economics. In this approach, we are most concerned with the details of the chemical process that we will use to make the chemically well defined commodity product. This concern with chemical process is one of the foundations of chemical engineering. It should not be abandoned even though the chemical industry now emphasizes specialty products, as opposed to commodity products. Chemical process concerns should be supplemented by the product concerns of this book; process concerns should not be disregarded.

A HIERARCHY OF PROCESS DESIGN

Here, we present only a précis of chemical process economics. This précis can serve as a review or as a counterpoint to the different ideas of net present value given in the next section. Chemical process economics can be conveniently organized with the design hierarchy mentioned earlier in Section 1.4. This hierarchy consists of four steps:

1. batch versus continuous;
2. input–output structure;
3. reactions, including recycles; and
4. separations and energy integration.

More elaborate hierarchies are easily imagined, but this one will give us enough for a good comparison.

The four steps implied by this hierarchy will usually be the intellectual sequence of any process design. We begin by deciding whether we will use a batch process or a continuous one. For most commodities, we will be making at least 1000 tons per year, which almost always implies a continuous process with dedicated equipment. We normally will never even consider batch equipment, which would be used for several different processes. We will go straight to the continuous process with dedicated equipment.

Our second step will be the input-output structure, most commonly in the form of a block diagram. One example is shown in Figure 7.2–1 for the manufacture of ammonia, the basic building block of agricultural chemicals. Not surprisingly, ammonia is made by combining nitrogen and hydrogen:

$$N_2 + 3H_2 \rightarrow 2NH_3. \tag{7.2–1}$$

Interestingly, the hydrogen is made from natural gas:

$$CH_4 + O_2 \rightarrow CO_2 + 2H_2. \tag{7.2–2}$$

Because the demand for anhydrous ammonia is seasonal, the carbon dioxide is often further reacted with ammonia to make urea and water:

$$2NH_3 + CO_2 \rightarrow CO(NH_2)_2 + H_2O. \tag{7.2–3}$$

The urea is a solid, and so is much more easily stored until spring than would be gaseous ammonia. Note that at this stage, our major intellectual tool is nothing more than reaction stoichiometry.

The third step begins to involve the details of the chemical reactions. These take two forms. First, the combination of nitrogen and hydrogen is a reaction that, in spite of catalysts, is either very slow or will not go to completion. We will

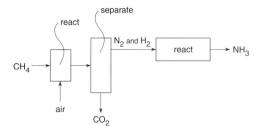

Figure 7.2–1. The Inputs and Outputs for Making Ammonia. At this stage, we need only the simplest block diagram, showing the chemical streams.

run this reaction hot, so that it will be fast. Unfortunately, this means that with a stoichiometric feed, we will only get about 20% conversion.

The combustion of methane to make hydrogen has the opposite problem. It goes to completion all too easily. Indeed, if we try to run the reaction shown, some of the hydrogen will burn to make water, the antithesis of a high value-added product. Instead, we will burn methane with much less air:

$$2CH_4 + O_2 \rightarrow 2CO + 4H_2. \tag{7.2-4}$$

We will then separate the carbon monoxide and run the water-gas shift reaction:

$$CO + H_2O \rightarrow CO_2 + H_2. \tag{7.2-5}$$

The results of these reactions will be a more elaborate process like that in Figure 7.2–2. In this effort, we have supplemented ideas of stoichiometry with concepts of reaction engineering.

Finally, in the fourth step of the design hierarchy, we need to specify the separations and the energy requirements. It is at this point that we will begin to use most aspects of chemical engineering. For example, we must decide how to separate the carbon dioxide and the hydrogen produced by the water-gas shift reaction. We can certainly do so cryogenically, by liquifying the CO_2, but it will be prohibitively expensive. We will probably choose to effect this separation by using an aqueous solution of a nonvolatile amine. We must decide whether the poorer capacity of a conventional amine, such as monoethanol amine, is better than the greater expense of a hindered amine, such as those described in Section 5.2.

As a second example of the complexity of this level of the hierarchy, we should remember that the air used to partially burn the methane contains not

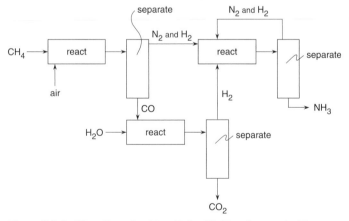

Figure 7.2–2. More Reaction Details for Making Ammonia. The combination of N_2 and H_2 is incomplete and so requires a recycle. The combustion of methane must use little air to avoid burning the hydrogen to make water. This produces carbon monoxide, which is reacted with water to make more hydrogen and carbon dioxide.

only oxygen and nitrogen but also argon. Although there is only about one percent of argon, there is no way for this noble gas to escape from the process in Figure 7.2–2. Most likely, we will need a purge stream on the gas recycled from the ammonia separation to the reactor making the ammonia. This purge will contain hydrogen, which is expensive; is there any way to recover at least the hydrogen from this purge? Questions such as this can be important in the production of commodities, where the value added and the profit margins are both small. The result is a considerably more complicated flow sheet, like that shown in Figure 7.2–3.

ECONOMIC POTENTIAL

When we are designing the process by using the hierarchy described above, we will make three sequential tests on the economic potential of the process. The first is to estimate the potential based only on the current prices of the product and the raw materials.

$$
\begin{aligned}
&[\text{economic potential (first estimate)}] \\
&= [\text{revenue from product sales per year}] \\
&\quad - [\text{raw material cost per year}].
\end{aligned} \tag{7.2–6}
$$

We would expect this potential to be positive for an attractive chemical process.

Three implications of this simple definition of economic potential are worth mentioning. First, the stoichiometry of the process will be important. For example, in the case of ammonia, air is presumably free, but methane certainly will not be. We will presumably make two moles of ammonia for every three moles of methane burned.

The second implication is similar: we must have finished steps 1 and 2 of the design hierarchy. In other words, we must have a reasonably exact idea of the process streams. Again, in the ammonia example, our estimate of economic potential will be dramatically altered by whether or not we run the water-gas shift reaction. If we do not, we get two moles of hydrogen from each mole of methane burned. If we do run this extra reaction, we can approach three moles of hydrogen from each mole of methane.

However, the truly dramatic implication of this simple definition of economic potential occurs when we try to apply it to both commodity products and to specialty products. For commodity products, it is easy: we can look up the current prices of ammonia and methane. Imagine that we try to do this for the shut-down battery separator, or the pollution preventing lithographic ink, or a new drug for treating depression. The battery separator sells for hundreds of dollars per kilo, but the polyolefin from which it is made sells for a few cents per kilo. The new ink costs less than the old one because it no longer uses a solvent; it is made from the same raw materials but with altered reaction conditions. The drug may be made from expensive raw materials, but its selling price is not really known. The price will depend not only on the drug's effects, but also on an elaborate cultural situation. Just think of Viagra, the drug that enhances penile erections.

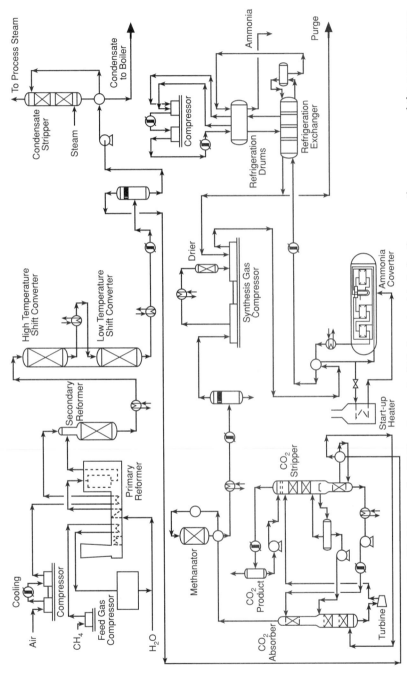

Figure 7.2–3. A Simplified Process for Making Ammonia. This flow sheet includes the separation processes and the process stream. Note also the purge stream for argon, which enters as an impurity in the air, and the heater required to start the ammonia reactor.

If the economic potential is positive after this first simple test, we can then apply a second criterion:

$$[\text{economic potential (second estimate)}]$$
$$= [\text{revenue from product per year}]$$
$$- [\text{raw material cost per year}] - [\text{utility cost per year}]. \qquad (7.2\text{--}7)$$

This implies that we know how much energy we will need to make our product. As a result, we must have finished at least initial estimates of all four steps of our design hierarchy. Again, to proceed we want this potential to be positive. If it is, we are ready to estimate how much the equipment will cost, and how much working capital we will need.

CAPITAL REQUIREMENTS

We now want to make our third estimate of economic potential, which includes our capital cost:

$$[\text{economic potential (third estimate)}]$$
$$= [\text{revenue from product per year}] - [\text{raw material cost per year}]$$
$$- [\text{utility cost per year}] - [\text{total capital cost per year}]. \qquad (7.2\text{--}8)$$

We need only to estimate the capital required in order to complete this economic précis.

Doing so is not difficult but complicated. There are no new ideas that are hard to grasp; there are just a large number of separate costs to keep track of. Moreover, to estimate each of these costs, we will use heuristics, rules of thumb that past experiences say are reasonable. Although these heuristics are normally reliable, they are affected by changing economic conditions such as inflation or the price of a barrel of crude oil. We should use them remembering that they are not scientific laws.

Our basic scheme for calculating capital requirements is shown in Figure 7.2–4. The scheme has five steps. In the first step, we must calculate the equipment costs. To do so, we use our process flow sheet, making estimates of the costs of

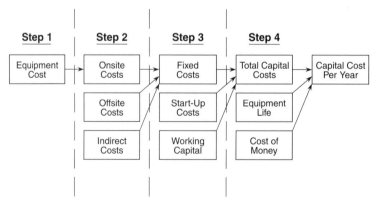

Figure 7.2–4. Estimating the Yearly Capital Cost. Knowing this quantity is the last step in estimating the economic potential of a commodity chemical process.

the different pieces of process equipment. For preliminary estimates, we can use tables and other guidelines listed in the literature, remembering to correct for inflation since the literature's publication. For later, more exact estimates, we will need to contact equipment vendors. With this knowledge of equipment costs, we can guess our onsite costs as

$$[\text{onsite costs}] = 4[\text{equipment costs}]. \qquad (7.2\text{--}9)$$

The factor of four in this relation is our first heuristic, an average of experience in the petrochemical industry. It may be larger if we are building a plant with used equipment. It can be smaller if our construction is in China, where labor costs are less. It will be much larger in specific industries, like nuclear power. In every case, we need to know this factor.

We now move to the second step of our scheme, and calculate the fixed cost:

$$[\text{fixed costs}] = [\text{onsite costs}] + [\text{offsite costs}] + [\text{indirect costs}]. \qquad (7.2\text{--}10)$$

The offsite costs include any power generations, such as steam or electricity, and any additional buildings, for management or technical support. As a general rule,

$$[\text{offsite costs}] = 0.45[\text{onsite costs}]. \qquad (7.2\text{--}11)$$

The indirect costs are dominated by in-house engineering and construction charges. From our experience, we expect

$$[\text{indirect costs}] = 0.25[(\text{onsite costs}) + (\text{offsite costs})]. \qquad (7.2\text{--}12)$$

Combining these relations, we find approximately

$$[\text{fixed costs}] = 1.8[\text{onsite costs}] = 7.2[\text{equipment costs}]. \qquad (7.2\text{--}13)$$

Remember the two heuristics with their factors of 0.45 and 0.25.

In the third step of our scheme, we calculate the total capital costs:

$$[\text{total capital costs}] = [\text{fixed costs}] + [\text{start-up costs}] + [\text{working capital}]. \qquad (7.2\text{--}14)$$

The start-up costs are for the extra engineers and operators whom we need to get the plant running:

$$[\text{start-up costs}] = 0.1[\text{fixed costs}]. \qquad (7.2\text{--}15)$$

The working capital includes the inventory of our product, where we have had to buy the raw materials but we have not yet sold what we have made:

$$[\text{working capital}] = 0.15[\text{total capital costs}]. \qquad (7.2\text{--}16)$$

Combining, we now have

$$[\text{total capital costs}] = 1.30[\text{fixed costs}] = 9.4[\text{equipment costs}]. \qquad (7.2\text{--}17)$$

Again, remember the factors of 0.10 and 0.15, introduced through two additional heuristics.

Finally, in the fourth step of this scheme, we estimate the capital cost per year from the total capital cost. To do so, we need the equipment life, often pessimistically assumed to be ten years. We also need the cost of money, which we will assume is 15%. Thus we find

$$[\text{capital cost per year}] = \frac{[\text{total capital cost}]}{10 \text{ yr}}(1.15)^{10} = \frac{4.0}{\text{yr}}[\text{equipment cost}].$$

$$(7.2\text{--}18)$$

This final result is used in Equation 7.2–8 to estimate the economic potential of this chemical process. If the potential is still positive, we may decide to proceed.

However, the point of this section of the book is not the details of this process design. The point of this section is that process design is different from product design. In chemical process design, we know that we have a market. We often know the prices in the market to a high degree of accuracy. In chemical process design, we expect to use continuously operated equipment dedicated to making one principal product. We must be concerned – indeed, obsessed – with the process details of making this product. For commodities, our profits may be huge, but our profit margins are a few percent.

This is almost never the case for specialty chemicals. For these, we are unsure whether a market exists. We may not know our product's price within a factor of two or more. We will sometimes make the product from relatively cheap raw materials. We will make batches, often in equipment used for different products. Think of these specialty products as coming from a microbrewery, not an oil refinery, and you begin to see why their design and their economics are so different.

One assumption made implicitly in all the above calculations for process economics is that we are operating essentially in steady state. In order to estimate whether a new plant is viable, we enquire whether it will make a profit on a yearly basis, taking into account the cost of capital, discounted over the lifetime of the equipment. In the case of chemical products we are usually in a very different situation. Chemical products can be expected to have a finite, often short, lifetime in the marketplace. In the case of chemical products we need to assess the economics of the project over the projected lifetime of the product. This is a very different proposition to deciding whether a plant turning out the same chemical year in year out will be viable. The next section outlines how the viability of a product over a short lifetime may be assessed.

7.3   Economics for Products

In our discussion of commodity chemical economics in the previous section, we assumed that we were selling into a pre-existing market and that the price of our product was determined essentially by existing competitors. Success in such a market implies producing a chemical at a lower marginal cost than the competition over a long period of time. The discussion of new product economics, by contrast, focuses on net present value, adjusted by the time to market and the time value of money. We neither assume that a market for our product already exists, nor that the price for our product is fixed by competitors already making this specific product. We imply that existing markets and competitors are not especially important.

What is important in determining a product's viability is its potential lifetime in the market. This is in stark contrast to commodities, where it is usually assumed that there will be a market for the product for as long as we produce it. The life of chemical products in the market may be limited by the expiration of patent protection, by the arrival of a competitor's better product, or by changes in fashion. Because these things are difficult to accurately predict, we usually assume that products will have a short lifetime. We will aim to turn in a healthy profit in five or at most ten years; any longer and the vagaries of the market will make our product too uncertain a proposition to pursue. The economics of products is focused on calculating the potential of a project over a short lifetime; fixed costs, notably research and development, must be written off over this short period.

Product economics can be summarized in various ways, including net present value, return on investment, and payback time. Net present value is simply the product's total value, in today's dollars, including all future cash flows. Return on investment is the total profit per year divided by the investment. The profit, which can be either the net or the gross value, does not normally reflect the time value of money. The payback time is the costs of fixed capital and start-up divided by the sum of yearly profit and the depreciation. The payback time can be harder to apply when the equipment is used in the manufacture of many products.

We will emphasize the net present value in this section. Estimating this value will often begin with a Gantt chart showing who works on the product. One example of such a chart is given in Figure 7.3–1 for a product with an expected four-year cycle. The bars show how deeply different groups are involved with the product as a function of product lifetime.

The figure has few surprises. The heaviest initial involvement is from marketing, who will normally be most involved in determining market needs. Research

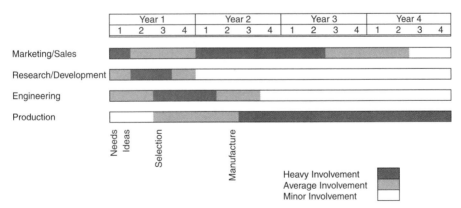

Figure 7.3–1. Gantt Chart Showing Project Involvement. The primary responsibility begins with marketing and then moves in turn to research, engineering, and production. Under the project system, however, all groups remain involved through the core team.

is also involved early, in establishing specifications, in generating ideas, and in resolving scientific issues which are important to the new product. Engineering will begin to be heavily involved during product selection, for that is when the detailed estimates of product performance and product manufacture will be central. Production will be most heavily involved during manufacture.

The core team for this product will all be heavily involved for the year and a half of product development. In this sense, the chart is misleading, because it de-emphasizes the importance of this team. The chart does so because it is reporting total involvement, including those on the core team. This total involvement is emphasized here because it reflects the total product design and development cost.

CASH FLOW WITHOUT THE TIME VALUE OF MONEY

We are now ready to discuss the cash flow for a typical product. We will do so first without considering the time value of money. We do this because we can see the ideas involved most easily. We will discuss the effect of the time value of money later in this section.

Our hypothetical example is shown in Table 7.3–1 for a product with a four-year lifetime. Note that the table implies development work for a year and new equipment in the second half of that year. Start-up costs occur in the fourth quarter of the first year and in the first quarter of the second year. Production costs also start in the fourth quarter and continue almost to the end of the project. Revenue begins in the second year, just as development ends.

The costs shown imply a wide variety of assumptions. The development costs assume a total of twenty persons. Each person paid an average of $60K, with 30% fringe benefits and 100% overhead, based on salaries alone. This overhead includes buildings, utilities, and management. This means that each person costs $138K per year, and that the total first year development cost is $2.76 million or $0.69 million per quarter. The total equipment cost is $2 million, spread over two quarters. The total start-up cost is twenty percent of the equipment cost, again spread over two quarters, but naturally after the equipment has arrived. Production costs $1 million per quarter for all but the last quarter. These production costs include those of sales and marketing. Revenue brings in $2 million per quarter, but not until after the first year. The lags in revenue are a reflection of inventory, or distribution and of collecting money owed.

The results of these costs and revenues are shown in the table. We spend $5.96 million before we see even a penny of revenue. We do not break even until half way through the third year. Although our total investment in development, equipment, and start-up is $5.16 million, our total net cash flow over four years is $6.84 million. Thus our return on investment is [($6.84 million/$5.16 million)/4 years], or about 33%. The net present value after four years is $6.84 million.

This simple example is especially helpful because it shows how important development costs are. These costs naturally occur early, and take a long time to pay off. The equipment costs also occur early, but tend to be smaller for product development than for process development. Finally, note how the cash flow jumps

TABLE 7.3–1 Cash Flow Without the Time Value of Money

Parameter	Year 1				Year 2				Year 3				Year 4			
	1	2	3	4	1	2	3	4	1	2	3	4	1	2	3	4
Development	−0.69	−0.69	−0.69	−0.69												
Equipment			−1.00	−1.00												
Start-up				−0.20	−0.20											
Production				−1.00	−1.00	−1.00	−1.00	−1.00	−1.00	−1.00	−1.00	−1.00	−1.00	−1.00	−1.00	
Revenue					2.00	2.00	2.00	2.00	2.00	2.00	2.00	2.00	2.00	2.00	2.00	2.00
Cash Flow	−0.69	−0.69	−1.69	−2.89	0.80	1.00	1.00	1.00	1.00	1.00	1.00	1.00	1.00	1.00	1.00	2.00
Net Cash Flow	−0.69	−1.38	−3.07	−5.96	−5.16	−4.16	−3.16	−2.16	−1.16	−0.16	0.84	1.84	2.84	3.84	4.84	6.84

Note: All values are in 10^6^. Note the time value to break even is in the third year and the return on investment is ($6.84/5.16/4) = 33%.

when, in the last quarter of the fourth year, we can sell off our inventory without production costs. We are, of course, incurring new costs for developing our replacement product. This simple example is also helpful because it illustrates how many factors we are ignoring. For example, we are assuming that the demand for our product starts as soon as it is available, and stays completely steady throughout the cycle. We are implying we can sell everything at the same price, without any reductions as we get to the end of the product cycle. We are ignoring any extra revenue that we might get by being first to market. We assume that any economic cycles have no effect, and that there is no inflation. We can imagine how all of these could be added to our financial analysis.

The most important factor we are neglecting is the time value of money. Accordingly, we include this factor in the paragraphs that follow.

CASH FLOW INCLUDING THE TIME VALUE OF MONEY

Imagine we want to buy a car costing $10K. If we buy it now, we need $10K. If we buy it in a year, we need less, because we can put the money in the bank, earn say 8% interest for the year, and buy the car at the end of the year. More specifically, we see that

$$[\text{money needed for car}](1.08) = \$10\text{K}, \qquad (7.3\text{--}1)$$

$$[\text{money needed for car}] = \$9.26\text{K}. \qquad (7.3\text{--}2)$$

Delayed gratification saves money.

The same effects, lumped together as the "time value of money," affect all the costs and revenues of product development. They change most of the numbers in an example like that in Table 7.3–1. The key effect is to alter the cash flows shown by means of the equation

$$[\text{adjusted cash flow}] = \frac{\text{present value}}{[1 + (\text{interest rate}/4 \text{ quarters})]^{n-1}}, \qquad (7.3\text{--}3)$$

where n is the number of quarters involved. Note that the adjusted cash flow equals the present value only in the first quarter of the first year, when n equals one. As an additional illustration, the adjusted cash flow for the $10K car mentioned above is

$$[\text{adjusted cash flow}] = \frac{\$10\text{K}}{[1 + (0.08/4)]^{4-1}} = \$9.42\text{K}. \qquad (7.3\text{--}4)$$

The cash needed is more than in our first calculation because, in this case, the interest will be paid back every quarter. We do not give them our money only once, for the entire year. Tom Wolf, describing differences such as these in his novel *The Bonfire of the Vanities*, calls them "the golden crumbs."

The cash flows in Table 7.3–1 are corrected for the time value of money in Table 7.3–2. In making these corrections, we assume that money costs 15%, or 3.75% per quarter. We can then correct the entries in the table. For example, for

TABLE 7.3–2 Cash Flow Including the Time Value of Money[a]

Parameter	Year 1				Year 2				Year 3				Year 4			
	1	2	3	4	1	2	3	4	1	2	3	4	1	2	3	4
Development	−0.69	−0.69	−0.69	−0.69												
Equipment			−1.00	−1.00												
Start-up				−0.20	−0.20											
Production				−1.00	−1.00	−1.00	−1.00	−1.00	−1.00	−1.00	−1.00	−1.00	−1.00	−1.00	−1.00	−1.00
Revenue					2.00	2.00	2.00	2.00	2.00	2.00	2.00	2.00	2.00	2.00	2.00	2.00
Cash Flow	−0.69	−0.69	−1.69	−2.89	0.80	1.00	1.00	1.00	1.00	1.00	1.00	1.00	1.00	1.00	1.00	2.00
Product Value[b]	−0.69	−0.67	−1.57	−2.59	0.69	0.83	0.80	0.77	0.74	0.72	0.69	0.67	0.64	0.62	0.60	1.15
Net Value[b]	−0.69	−1.36	−2.93	−5.51	−4.82	−3.99	−3.19	−2.42	−1.67	−0.95	−0.26	0.41	1.05	1.67	2.27	3.42

[a] As in Table 7.3–1, all values are in 10^6. The difference between this and the earlier table is that cash flow is adjusted for 15% interest. The time to break even is a quarter longer, and the net present value drops 50% from $6.84 million to $3.42 million.
[b] Based on the first quarter of Year 1.

the first quarter of the second year, our total costs are ($0.20 − $1.00 + $2.00 = $0.8) million. But because we do not see these costs for over a year,

$$[\text{adjusted cash flow}] = \frac{\$0.80\,\text{million}}{[1 + (0.15/4)]^{5-1}} = \$0.69. \tag{7.3-5}$$

The value is reduced by being in the future.

These corrections for the time value of money significantly alter the net present value. Without corrections, the value given in Table 7.3–1 is $6.84 million. With these corrections, the value given in Table 7.3–2 is $3.42 million, a reduction of 50%. This is without all the other concerns listed above, such as variable product demand and inflation. Every time we review this example, our respect increases for those in business who tame these numbers to make profits. We re-emphasize that such number taming is essential to product development, even though it does not center on engineering skills.

TIME TO MARKET

These concerns with cash flow neglect time to market, the other unusual factor in product economics. The business cliché is that the first company to bring a new product to market gets 60–70% of the total sales of the product. Competitors who arrive late fight over the remaining fraction.

This cliché seems to frequently be true for chemical products. For example, the Du Pont product Nafion is a perfluoronated ion exchange film used to separate the electrodes in the chloralkali process. Nafion, an enormous improvement over earlier technology, has become the best choice for manufacturing sodium hydroxide and chlorine. Because Nafion is no longer completely covered by patents, both Dow and Asahi Glass now make competing products that are by some significant measures superior. However, as the first of these similar products, Nafion continues to dominate sales.

There are well known exceptions to this first-to-market cliché. For example, ultrasonic imaging was developed for medical uses by an English company. When General Electric realized that ultrasonic technology threatened their large X-ray business, they mounted a crash program in the ultrasonic imaging business. Though they began from extreme weakness, they did succeed in capturing a dominant market position by emphasizing product quality. Although General Electric was not first to market, they eventually captured a majority share.

In a similar scenario, Eli Lilly found that their porcine insulin business was suddenly threatened by Nova, a Danish company, which had succeeded in making commercial quantities of human insulin by using the methods of genetic engineering. Eli Lilly developed a successful crash program to produce human insulin. They succeeded, protecting their dominant market share of this large business.

The main reasons why being first to market may not always be most important concern product quality and product cost. Again, a drug such as insulin is a good example. The company who can first get approval and bring a drug to market will certainly get most of the initial sales. If the drug can be made at higher purity,

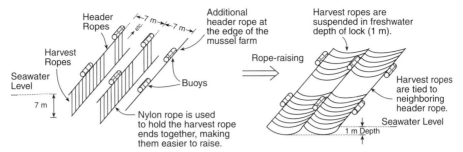

Figure 7.3–2. Mussel Harvest Ropes. As explained in the text, these ropes allow raising the mussels from salt water into fresh water, thus reducing the tube worms attached to the mussel shells.

or at least more free of pyrogens, then a second company has a good chance of gaining market share at the expense of the first. If a third company can get both high quality and lower cost, they have a chance, especially when the drug comes off patent. Nonetheless, we believe that the business cliché will normally hold, and that the first to bring a new product to market will get a clear majority of the product sales. In product design, be quick.

EXAMPLE 7.3–1   THE ECONOMICS OF SCOTTISH MUSSEL FARMING

Mussels are farmed on vertically suspended ropes in Scottish lochs. One problem for mussel farmers is that tube worms settle on the mussel shells and leave a hard, calcinaceous deposit. Although they leave the quality of the mussel meat unaffected, the worm casts are considered unappealing by consumers and in particular restaurants, which account for a significant fraction of mussel sales. Farmers are forced to manually sort and discard seriously fouled mussels to maintain the product quality. Typically 5–25% of the mussel crop is lost in this way.

One idea for mitigating this loss comes from the nature of Scottish lochs, which are narrow fingers of sea water reaching inland. Although the bulk of the water is salty, a significant quantity of fresh water flows into the lochs. Because of its density difference, the fresh water floats, resulting in a salt depleted surface layer of 1 m depth. It turns out that tube worms are less tolerant to fresh water than mussels. It should therefore be possible to discourage tube worm settlement and growth by periodically raising the mussel ropes into the top 1 m of water. We plan to do this by pulling the vertical ropes into a near horizontal position, as shown in Figure 7.3–2.

Investigate the economic viability of this project.

SOLUTION

In current practice, mussel harvest ropes are 7 m long and 30 cm apart, hanging from long horizontal header ropes. We will modify the current network of ropes in a mussel farm by attaching nylon ropes (which run parallel to the main header rope) to the ends of the dangling harvest ropes, thus easing the process of raising a number of these ropes into the required horizontal position. We decide on the following specifications:

1. Harvest ropes are grouped in batches of ten.
2. Nylon ropes of 32 mm diameter and 3 m length will be tied to the ends of ten harvest ropes to form a sideways ladder "header" rope.
3. The header ropes will be anchored such that they run parallel at an approximate distance apart of 7 m. An extra header rope (with no harvest ropes attached to it) will run parallel to the others at the edge of the mussel field to accommodate the harvest ropes.

A typical mussel farm has 15,000 harvest ropes, producing 150 tons of mussels annually, which sell for £810 per ton. We require 3 m extra rope per ten harvest ropes (i.e., 4500 m additional rope in total) to implement our solution. Because 200 m of 32-mm nylon rope costs £550, we will have a cost of extra rope of £24,750. The ropes will need to be raised every two weeks for a couple of days during the tubeworm breeding season (May to August). This takes about 32 man days of labor, or around £3200 per year. (Casual labor is cheap in Scotland.)

A pilot study shows that raising the ropes achieves a 10% increase in mussel yield per year. This gives an increase in income of $150 \times 0.1 \times £810$ or £12,150 per year. Finally, we expect that the ropes will need replacing after 5 years.

We are now in a position to calculate the net present value of the project over its 5-year life cycle. We assume that the cost of money is 8% per year. To begin, we see that

$$\text{Net income gain per year} = £12{,}150 - 3200 = £8950.$$

Though capital is spent at the start, income is gained in each of the successive five years, before the ropes must be renewed. Thus

$$\text{Net present value} = -24{,}750 + \frac{8950}{1.08} + \frac{8950}{1.08^2} + \frac{8950}{1.08^3} + \frac{8950}{1.08^4} + \frac{8950}{1.08^5}.$$
$$= £10{,}985.$$

We expect the investment to pay off, with a healthy return on investment of over 40%. The profit will be affected by factors such as the increased yield of mussels and the labor required to raise the ropes.

7.4   Conclusions and the Fourth Gate

The economic précis in this chapter emphasizes the differences between the key costs of commodity processes and specialty products. Because commodity products are usually needed in larger amounts, they are synthesized in dedicated equipment, operated continuously to spread out its huge capital cost. Because specialty products are made in small amounts, they tend to be made in batch equipment, which is used for several different products and may already be fully depreciated.

The result is two different approaches for evaluating product value. For large-scale commodity products, we will often begin estimating our economic advantage as the value of our products minus the cost of our raw materials. Note that this assumes we can accurately estimate our product's selling price. We then calculate our other costs, especially the capital costs of chemical process equipment. The costs of chemical equipment are first estimated by using simple heuristics. Later, these costs can be refined with more accurate computer-aided process design, and with detailed conversations with vendors. Usually the high capital costs involved in constructing dedicated process equipment can only be justified by continuous production over a long time scale; we assume that a long term market exists for our commodity chemical.

For small-scale products, we will normally find ideas like net present value are useful. In these cases, we will usually have products of high added value, where raw material costs are not central. We will focus instead on the costs and the speed of product development, for time to market is critical. We will also require an attractive return over a relatively short period of time, as a product's lifetime in the marketplace is unpredictable, and often short.

Although the economics outlined in this chapter can be a useful start, they are a pale imitation of those required. Whether we plan to make commodity chemicals or chemical products, we will need much more detailed estimates, which include many more factors in addition to those discussed here. These estimates will be interactive, requiring skills and resources that most chemical engineers do not have. They will be the responsibility of those with stronger backgrounds in business. Engineers and scientists are more necessary to the aspects of product design described earlier in this book.

This point in product design is the last big management review, the fourth of the four "gates" through which the project must pass. As a result, we can benefit from a review of the earlier gates as a summary of the obstacles that we must overcome. The first gate, after we had defined needs and specifications, tends to be superficial. This is partly because the project is so new, and partly because the reviewers – the management team – often have a role in defining the need. At this point, we with technical training probably need to be sceptical, damping the managers' enthusiasm.

The second gate, a review of our screened ideas, is also benign. Our managers may be charmed by what science can offer and intrigued by our potential inventions. Again, because the project is new, they will be tolerant. At this point, we should be critical, even pessimistic, about our own ideas. We need to do so especially because many studies suggest that few product design efforts are stopped too early, and that many are allowed to continue too long.

The third gate comes after selection, when we have chosen our best new product. At this point, our management will become much more engaged, even hostile, because we are asking for a lot of money. Management may or may not understand chemistry and engineering, but they know money well. Now, our job is neither to be sceptical or critical nor optimistic; our job is to be as accurate as possible. We and our organization are about to take a significant risk.

Finally, at the fourth gate, we will know how we want to manufacture the product. At this point, the management should have become our allies, not just our critics. After all, they, too, will want the new product to be successful. This means that they are less likely to be as critical as they were at the third gate. This in turn means that we need to have our technology down pat. No one will give us partial credit if we get the wrong answer, even though we had the right idea. There is no partial credit in the fascinating game of chemical product design.

FURTHER READING

Douglas, J. M. (1988). *Conceptual Design of Chemical Processes*. McGraw-Hill, New York, ISBN 0070177627.

Hall, D., Jones, R., and Raffo, C. (1993). *Business Studies*. Causeway, Ormskirk, ISBN 1873929099.

Peters, M. S. and Timmerhaus, K. D. (1991). *Plant Design and Economics for Chemical Engineers*. McGraw-Hill, New York, ISBN 0070496137.

Turton, R., Bailie, R. C., Whiting, W. B., and Shaeiwitz, J. A. (1998). *Analysis, Synthesis, and Design of Chemical Processes*. Prentice-Hall, Upper Saddle River, NJ, ISBN 0135705657.

Ulrich, K. T. and Eppinger, S. D. (2000). *Product Design and Development*, 2nd ed. McGraw-Hill, New York, ISBN 0071169938.

Problems

CHAPTER 2. NEEDS

2.1 **New Product Ideas.** Go into a foodstore, a drugstore, or a hardware store to identify three new potential products. Describe in a sentence or two what need each new product would fill.

2.2 **Reusable Detergents.** A summary of the raw materials and final uses of various detergents is given in Figure P2.2. Which current applications do you think are most promising for initial efforts to develop reusable detergents?

2.3 **Garden Tools.** Your management has suggested that the handles of garden tools – shovels, hoes, picks, and so on – are a good market for composite plastics. Using wood as a benchmark, set specifications that any composite materials must exceed. Among the factors that you may wish to consider are strength, weight, and price.

2.4 **Using Rainwater.** In order to avoid the rapidly increasing water charges made by utility companies and to ensure a completely pure supply of domestic water, we want to design a system that makes a typical family home self-sufficient in water. Rain water will be collected from the roof, and this must supply all the house's water needs. List needs and rank them in order of importance. Quantify these needs.

2.5 **Better Dental Fillings.** Many adults in developed countries have "fillings" in their teeth. These gray-colored amalgam insertions repair regions where the original hydroxyapatite, a calcium phosphate, has decayed. The fillings, often acquired during adolescence, last decades.

Recently, the public has reacted favorably to tooth-colored fillings, or as dentists call them, restorations. These newer restorations are often ceramics, which can be colored and wear well. However, the ceramics' hardness means that they can fracture easily: they don't absorb shock well. In addition, they are so hard that they cause the bone to which they are attached to demineralize.

As a result, manufacturers have developed systems based on polymeric resins, especially acrylics. Originally, these resins behaved very differently than the amalgams they replaced, so dentists had trouble using them. Now, resins are made to handle like amalgams. Also, the original resin systems wore too rapidly, even when filled with 50-μm silica. Now, resin systems use 0.06-μm silica for wear and 0.8-μm silica to adjust color.

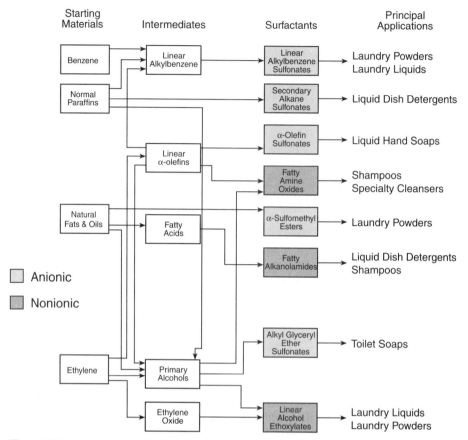

Figure P2.2.

Your company manufactures a variety of inorganic colloids of controlled size and shape. You are interested in exploring the use of your products in dental materials. How should you begin to explore this possible use?

2.6 **Degumming Soybean Oil.** Soybeans supply the standard cooking oil in North America. The beans are crushed and extracted with hexane. The spent beans are then leached with base, producing an aqueous protein solution and a high fiber residue. The proteins are separated from water and used as a food supplement. For example, soy protein is a low fat, low cholesterol egg substitute in commercial baking. The high fiber residue is used in health foods.

The oil-containing hexane extract is currently washed with water to remove "gums" and then washed again with caustic to remove fatty acids. These two steps produce a dilute, low value waste stream that can be difficult to sewer. As a result, there is significant interest in alternative methods to "degum" this extract.

One possible technology is the membrane process of ultrafiltration, which could remove the gums to produce an oil-hexane solution easily separated by distillation. However, ultrafiltration must overcome two major barriers:

(1) The ultrafiltration membranes must remain viable in the hexane, which is not now the case.

(2) The gums must not severely foul the membrane.

Note that ultrafiltration is currently successful in separating proteins from cheese whey, a roughly similar but aqueous separation.

Using the cheese whey case as a benchmark, set tentative specifications for the ultrafiltration of soybean oil extracts.

2.7 Biodegradable Plastics. There is considerable interest in producing biodegradable plastics from renewable feedstocks. What are the major markets for such a product likely to be? Who are the customers? List the major needs that it will be necessary to satisfy if such products are to succeed. Suggest a benchmark.

2.8 The Perfect Margarita. Doesn't it really irritate you when you get back to the veranda after a hard day on the beach and mix yourself a cool, refreshing cocktail, only to find it has been diluted beyond recognition by the melting ice? This problem must be solved. It is proposed to manufacture a device to produce the perfect margarita from ingredients at ambient temperature. Draw up specifications for what this device must be able to achieve. Be as quantitative as possible.

2.9 Global Warming. It is now widely accepted that increasing carbon dioxide levels in the atmosphere are causing warming of the planet by preventing the escape of infrared radiation into outer space. This is a kinetic problem rather than an equilibrium one. Carbon dioxide from the atmosphere dissolves in the sea and is then deposited as limestone, either by precipitation or via secretion into animal shells. The capacity of the sea to hold carbon dioxide is vastly larger than the amount of carbon held in reserves of fossil fuels, and the amount of carbon locked away in limestone is many thousands of times larger again. Projects are under consideration for speeding the transfer of carbon dioxide from the atmosphere to the oceans and rocks. Suggest criteria that should be used to assess these projects and rank these criteria.

CHAPTER 3. IDEAS

3.1 Teeth Whitening. Teeth bleaching, practiced by dentists for over a century, normally uses a 35% hydrogen peroxide solution to whiten the teeth. Because teeth are only 85% mineral by volume, the bleaching reaction is not only at the surface. The chemistry involved is unknown, because the causes of tooth color are uncertain. Nonetheless, the typical bleaching in the dentist's office takes one 45-minute visit per week for a month.

Your company makes a product for whitening teeth, which is essentially an aqueous solution of 10% carbomide peroxide. This is felt to be equivalent to a 3% solution of hydrogen peroxide. This product is worn in a special tray around the teeth for two 1-hour treatments per day. While it works, the length of the treatment limits the number of customers.

How can you make your product work more quickly?

3.2 Nanoreactors. Progress in microelectronics has included development of etching techniques capable of making very small structures. These small structures have simulated interest in nanotechnology: nanovalves, nanoturbines, and so on. This example is concerned with nanosized chemical reactors.

We can immediately think of several advantages that these reactors might have. To begin, they could provide very rapid mixing. However, turbulent mixing can produce eddies around 30 μm. In gases, these mix in microseconds; even in liquids, they mix in seconds. Thus fast mixing alone would not seem to be a good reason for these nanoreactors.

Traditional Tester
Bulky. Too many parts.
Relatively complicated.

On-Package Tester
Compact. Easy to use.
Also easy to lose.

On-Battery Tester
Tester right on battery.
Impossible to lose.

Figure P3.3.

The reactors do offer potential advantages in terms of selectivity. If their walls are coated with catalyst, they provide a single, well-defined residence time at a single temperature. Dispersion caused by diffusion within porous catalysts is eliminated. This should be equally true for reactions in gases or in liquids.

Suggest systems where these reactors might be valuable.

3.3   Better Batteries. Eveready is one of the world's largest manufacturers of batteries, and the company who introduced the battery tester attached directly to the battery. Their summary of this product feature, taken from their web page, is shown in Figure P3.3.

If you work for a competing company, what would you do to improve this technology even more?

3.4   Reliable Birth Control. Your company manufactures birth control products. Market research has shown that rather than take a pill every day, many women would prefer a product that automatically releases hormone into their bodies in the correct dose and lasts for a period of 6 months. List ideas as to how this might be achieved.

3.5   Safe Smartcards. Smartcard processors are used in applications such as direct debit cards, phone cards, and chips for decoding satellite TV channels. Breaking into these chips, extracting both their hardware and software, and reproducing this on huge numbers of fakes is big business. In some countries, such as Germany and Canada, this procedure is not even illegal. Multimillion dollar operations, complete with the most modern technology and some extremely bright scientists, exist solely for the purpose of cracking the codes of smartcard processors. The challenge is to design a way of making it extremely difficult to break into the relevant chip without destroying the information held within it. Brainstorm to produce a list of ideas for ways to tackle this problem. Prune and sort this list.

3.6   Detergent tablets. In Europe, the current trend in laundry detergents is to produce them in the form of a tablet, rather than a free flowing powder. This removes the need to measure the quantity to be added, avoids spillage of powder, and reduces clogging of dispensing funnels with wet powder. (In contrast to most U.S. machines, European washing machines dispense powder through a small funnel at the front.) In order to prevent these tablets from absorbing water and sticking together, each tablet has to be wrapped in plastic. This is expensive and also requires removal before use, which results in unpleasant handling of the detergent. Suggest alternatives to overcome this problem.

TABLE P3.9   Alternatives to Insulin Injection

1. Inject the insulin without needles.
2. Attach it to another molecule, so you can take it as a pill.
3. Fix this by genetic engineering.
4. Make a slow release form, so you need to inject less often.
5. Build a surgically implanted pump, which releases insulin when needed.
6. Implant modified insulin, which dissolves only when glucose is present.
7. Use another drug.
8. Make the implantable pump very small.
9. Push insulin across the skin with an electric field (this is sometimes called "electroporation").
10. Modify insulin so it can diffuse across the skin.
11. Have an insulin patch, like a nicotine patch.
12. Give insulin as an aerosol, and sniff it.
13. Build a surgically implanted machine that gives off insulin pulses.
14. Put insulin in vesicles, little microcapsules with an insulin center.
15. Build an artificial pancreas.
16. Use a mixture of different drugs.
17. Inject microspheres with attached insulin.
18. Use many very small needles.
19. Use no needles.
20. Make an insulin reservoir and a valve that the patient can open when he or she needs to.

3.7   Faster Chekouts. Consider how much time and effort is wasted at the supermarket checkout in individually passing items past a scanner in order to record their price. Imagine how much better it would be if a full grocery cart could be wheeled through the checkout and the list and price of all the items automatically compiled. How might this be achieved?

3.8   Clean Feet. Foot odor is a major cause of romantic failures. List and sort possible solutions.

3.9   Insulin Aerosols. Diabetes is a disease in which the body is unable to metabolize sugars and carbohydrates because the pancreas does not secrete sufficient insulin, a small protein (mol. wt. 5700). Patients often treat this disease by measuring their blood glucose; when it is high, they inject themselves with shots of insulin. The injections are necessary because insulin can't survive in the stomach's acid. Even if it could survive, it is not transported across the wall of the intestine.

Your company is interested in a product that would reduce the need for these insulin shots. As part of "idea generation" for such a product, your core team comes up with the concepts in Table P3.9. (These are actually a fraction of the total.)

(a) Organize these as part of your "idea sorting." In this organization, refer to specific ideas.

(b) Choose three ideas that you feel are especially good and construct a decision matrix to illustrate your rationale for preferring one of these. Briefly justify your scoring.

3.10 Wine Aeration. The ideas from a brainstorming session seeking a product that will improve red wine by aeration are given in Table P3.10. Organize these ideas, and decide which are most attractive. (One set of solutions is given as Example 5.3–2.)

TABLE P3.10 Improving Wine Breathing

1. Better if poured three times.
2. Want to add oxygen.
3. Want some entrained air (not too much).
4. This only works for red wine.
5. Use a "ship's decanter."
6. Use a compact falling film evaporator.
7. Same effect works for green tea.
8. Pouring gives better flavor but less bouquet.
9. Better taste means less acid.
10. Polyphenols react with oxygen to improve flavor.
11. Use electrolysis to make oxygen.
12. It must depend on the kind of red wine.
13. Put the wine in a blender.
14. We need an oxygen sensor.
15. Use a fish bowl aerator.
16. Will it work better if hot?
17. We could use an oxygen sensor.
18. The kinetics of flavor are faster than making vinegar.
19. Use structured packing for gas absorption.
20. Add a pill containing iron filings.
21. Use selective adsorption.
22. Does it depend on the size of bottles?
23. Add oxygen with a syringe.
24. Add oxygen with a straw.
25. Add a pill that gives off oxygen.
26. Add bamboo leaves (they work in green tea).
27. Use a selective adsorbent.
28. Is taste or smell more important?
29. Stir the wine magnetically, so we get a vortex.
30. Is the market for making cheap or expensive wine better?
31. Use hollow-fiber membranes.
32. Aerate with N_2, not O_2.
33. Aerate with CO_2.
34. Do winos have the same preferences as gourmets?
35. In China, wine is served hot.
36. In Japan, sake is served hot.
37. The machine for wine breathing should be in the wine store.
38. No, it should be at home; that's a bigger market.
39. Aerated wine tastes "round."
40. Oxygen should come from the cork, which could be a special plastic.

CHAPTER 4. SELECTION

4.1 Comparing Fruit. Construct a selection matrix to compare the merits of apples and oranges.

4.2 Glass Restoration. King's College Chapel at Cambridge University has the largest surviving area of fifteenth century Flemish glass in the world. Instructions for repairing this glass that have survived use human urine as an ingredient. Urine is not a standard chemical preparation. Suggest what should be used as a substitute.

4.3 Blood Bags. Blood and physiological saline are currently stored in one-liter plastic bags, a familiar feature of hospitals and paramedical units. These bags are usually made of polyvinylchloride. Because this polymer is hard at room temperature, it is "plasticized" (made flexible) with the compound di(2-ethylhexyl)phthalate. Unfortunately, the compound is carcinogenic and leaches into the bags' contents over time. The Environmental Protection Agency's admitably arbitrary limit is 6 ppb. One bag manufacturer measured the concentration in current saline as 5000 ppb.

Outline how you would choose a new material from which the bag would be made. (*C & E News*, March 15, 1999.)

4.4 Warm Seats. You want to test two possible ideas for a cushion to keep your seat warm during a three-hour sports event. The first idea is a pad containing rechargeable batteries; the second is for a pad filled with an inorganic hydrate that freezes at a comfortable temperature, releasing its heat of fusion.

Which is the better idea? To make your comparison quantitative, assume you are sitting on a 5-cm plank in 5°C air, and that you want to keep your seat at 25°C.

4.5 Espresso Coffee. One rapidly growing specialty product is prepackaged coffee for individual espresso machines. The product is a metal foil bag containing exactly the right amount of specially prepared coffee to make a single cup. Foil is used to keep out oxygen, which radically compromises the coffee's flavor. The bag fits directly into the machine, which punctures it to make the coffee.

The problem is that freshly roasted and ground coffee contains a lot of carbon dioxide, which is released over time. In some cases, this released gas bursts the foil bag. You are considering two possible cures for this problem:

(1) A foil bag that has a one-way valve to release the extra pressure.
(2) A transparent polymer bag that is permeable to carbon dioxide but not to oxygen.

Describe how you would choose between these two alternatives.

4.6 Coolers for Oil Platforms. The size of oil and gas exploration platforms is extremely sensitive to the size and weight of the equipment to be supported above sea level. Key components of gas production platforms are the gas compressors used to transport the product to shore; their efficiency and size is optimized by the use of intercooled multistage units. The coolers, however, are often larger than the compressors – one prevalent existing technology is to use banks of fin-fan coolers (air/gas or air/liquid coolant), which are limited by the maximum obtainable rates of heat transfer to the surrounding air. More efficient cooling solutions are under investigations. The following are the most promising three alternatives to fin-fan coolers:

(i) Packed bed cooling. The hot gas is countercurrently contacted with pure, cool water. The pure water is then cooled by seawater.
(ii) Direct injection of liquid coolant. Latent heat and sensible heat are transferred to the added liquid, thus cooling the natural gas.
(iii) Heat pipes. Heat is transferred from the evaporation section of a heat pipe to the condenser section by using the equilibrium between liquid and vapor water under vacuum conditions.
(iv) No cooling. Just lag the pipeline and let it cool as it is piped away from the platform under the sea.

Assess these ideas relative to the benchmark of fin-fan cooling. Assume natural gas enters the cooler at 10^4 kPa 170°C and leaves at 90°C. About 100 to 250 million ft^3 of gas must be cooled per day.

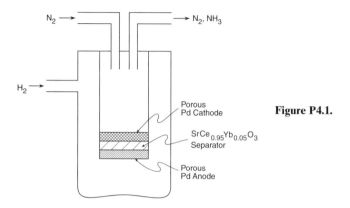

Figure P4.1.

4.7   A Mini-Haber Process. Ammonia is largely made by the Haber process, an early triumph of industrial chemistry. This process burns natural gas to produce a mixture of nitrogen, hydrogen, and carbon dioxide. After the carbon dioxide is removed by amine scrubbing, the nitrogen and hydrogen are reacted at $450°C$ and 200 atm over an iron catalyst to produce some ammonia. Argon in the product must be purged; unreacted N_2 and H_2 are recycled.

Although the Haber process has been optimized by almost a century of effort, the extreme reaction conditions require large capital investment in large plants, and a complex distribution system to deliver the ammonia to farmers. As a result, your company is interested in developing a small ammonia plant for individual farms. Your development team has decided to focus on two technologies. The first is the Haber process itself, scaled down and carried out in high temperature, high pressure microreactors. The second is a high temperature, atmospheric pressure electrochemical process (*Science*, 1998, **282**, 98–100).

Some details of this second process are shown in Figure P4.1. The process passes hydrogen across a nickel anode to make protons. These are electrochemically driven across an inorganic membrane separator at $570°C$ to react at a palladium cathode to produce ammonia. The reaction is 78% efficient; 22% of the protons are lost as heat. Still, a 78% conversion is much better than the Haber process, in which conversion is typically around 20%.

Set up a selection matrix to help decide if you want to pursue development of either of these processes.

4.8   Better Orange Juice. Fresh orange juice tastes much better than juice made from frozen concentrate. There are two main reasons for this. First, orange juice flavors, exemplified by octanal, tend to decay with time, especially in the presence of oxygen. Second, orange juice concentrate is produced by vacuum evaporation, in which volatile flavors are lost. To replace these lost flavors, some manufacturers add ground orange peel (in standard products) or unevaporated fresh juice (in premium products).

Your company wants to investigate alternative processes to produce a premium juice concentrate. After a brief technology review, you decide that distillation, freeze drying and reverse osmosis are the best alternatives to your benchmark, vacuum evaporation. You decide that the key criteria for choosing between these processes are thermal degradation, volatile loss, engineering difficulty, and product cost.

Prepare a selection matrix to help you decide which of these technologies merits further development. Briefly justify the weighting factors that you use in this matrix. Finally, choose one technology and make a risk analysis.

4.9 **Pipeline Plugging.** Wax deposition in subsea pipelines that are used for the transport of crude oil can cause flow reductions or even blockages. This problem has been estimated to cost oil companies $100 million annually. Our main challenge in clearing such pipeline blockages is in supplying heat to downstream regions of the pipeline in order to melt the wax, which is then dispersed by using injected surfactants. In subsea pipelines, the oil is continually cooled by the sea at $4°C$ and blockages may be tens of kilometers downstream of the point of injection. We must provide the heat selectively at the point of blockage. We propose to achieve this by using a fused exotheric chemical reaction, that is, one with a built in, controlled delay before commencing. The reaction proposed is:

$$NH_4Cl_2(aq) + NH_4NO_2(aq) \rightarrow 2H_2O + NaCl(aq) + N_2(g),$$
$$\Delta H = -334.5 \text{ kJ mol}^{-1}.$$

The reaction is catalyzed by acid.

Two methods of achieving and controlling the time delay are being suggested:

(i) Encapsulation of the reactants and catalyst in separate capsules that then disintegrate over a controlled period of time. The reaction is initiated when the catalyst mixes with the reactants.

(ii) Pulses of reactants and catalyst are separated by an inert spacer – contacting then occurs as a consequence of dispersion as the material travels down the tube.

Consider criteria on which to judge the methods and so justify selection of one of them. Suggest information that would be needed before manufacture of the product could proceed.

(Singh, P. and Fogler, H. S., 1998, Fused Chemical Reactions: The Use of Dispersion to Delay Reaction Time in Tubular Reactors. *Ind. Eng. Chem. Res.* **37**, 2203–2207; Subramanian, K. and Fogler, H. S., 1999, Fused Chemical Reactions: Use of Encapsulation to Delay Reaction Time. *Proc. 4th Italian Conf. on Process Engineering*, Florence, 43–46.)

4.10 **Antidepressants.** The patent for Prozac, a major antidepressant, will expire in 2003. As a result, there is interest in a generic version of this drug, whose active ingredient is fluoxetine. A review of the literature suggested two attractive chemical syntheses, an enzymatic synthesis and a fermentation. Key references follow:

(a) A synthesis with mondelic acid: Koenig and Mitchell (1994), *Tetrahedron Lett.* **35**, 1339.

(b) A synthesis with diisopinocampheylchloroborane: Braeher and Lity (1996), *Bioorg. Med. Chem.* **4**, 877.

(c) The enzyme lipase: Scheider and Goergens (1992), *Tetrahedron: Asymmetry* **3**, 525.

(d) A fermentation using *Beauveria sulfurescens*: Chenevert et al. (1990), *Tetrahedron* **48**, 6769.

Read these references. Then develop a selection matrix that compares these alternatives based on the efficiency of the synthesis, the cost of raw materials, and the cost of equipment.

4.11 Bacteriophages. Many bacteria show increasing resistance to penicillins. This is largely caused by overprescription in the West. A possible solution is the use of bacteriophages. Discovered independently by a Canadian biologist, d'Herelle, and an English microbiologist, Twort, bacteriophages are a class of virus that "eat" bacteria. There are many types of bacteriophage, each of which attacks a specific bacterium. Bacteriophages found limited use in treating dysentery in the First World War but were never used widely in Western medicine, and they completely disappeared following the discovery of penicillin. However, their development in the Soviet Union continued, particularly in Georgia. In Tbilisi, bacteriophages have been used on a large scale and to treat a wide variety of diseases very successfully over decades. However, the Soviet documentation does not meet FDA standards and the scientific resources in the Tbilisi labs are primitive by Western standards.

Recently there have been a few high profile cases of successful use in the West, where the alternative was certain death. Your company proposes to develop bacteriophages in collaboration with Tbilisi scientists for use on penicillin-resistant bacteria. Consider the risks involved in this project.

CHAPTER 5. PRODUCT MANUFACTURE

5.1 Practice in Experiment Design. At this point in product design you will often need to supplement your calculations with some simple experiments, just to make sure that you are on the right track. You may have trouble designing these simple experiments. To practice, try making one or more of the following experiments with a ruler, a watch, a thermometer and a scale:

(a) Measure the drying rate of a wet gym towel and compare this with that estimated from mass transfer correlations.

(b) Estimate the changing strength of coffee coming out of the coffee grounds as the pot is brewed.

(c) Measure how long your coffee takes to cool and compare this with predictions of heat transfer correlations.

In every case, connect your quantitative measurements with what you know about engineering. (Hans Wesselingh, University of Groningen.)

5.2 Hot Sake. The following are unedited translations of Japanese originals. The first is from the web page of one of the biggest sake producing companies.

> "Warmed or slightly heated sake is called 'kan'. When kan is served at 45°C (113°F), its fullness of body and mellow flavor become more pronounced making this a popular choice during the cooler months or when paired with refreshingly light fare. Enjoyed in this fashion, kan is particularly soothing."

The second translation is from the label of a pull-tab can of sake (shown in Figure P5.2):

> "Kan-Ban Musume (This has a double meaning: the most pretty waitress in a bar, and a girl in charge of warming sake in a bar). Amount: 180mL; Alcohol content: 14–15%; Price: ￥290($2.65)" (Simple canned sake is sold at ￥230)

> "How to warm sake in can?"

>> "1. Put can with facing up the plastic cap on bottom. Remove the cap and push the top center strongly until the top plate concaved. Keep pushing for another 5 seconds.

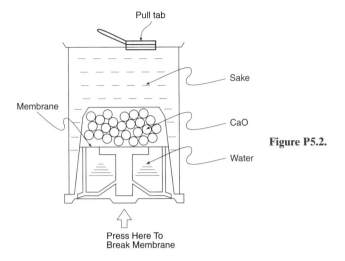

Pull tab

Sake

Membrane

CaO

Figure P5.2.

Water

Press Here To
Break Membrane

2. Put back the plastic cap on the bottom and fix it tightly.
3. Turn over the can and touch the upper side to feel the heat coming from the bottom. If you feel the heat, just pull the pull-tab up to release the pressurized gas inside and wait for a couple of minutes. If the heat doesn't come, go back to the 1st step and repeat again.
4. After 3 or 5 minutes, the temperature of the sake will reach 35–40°C. Now enjoy your warm sake."

By pushing the bottom of the can the water in the container is brought into the bed of calcium oxide and it triggers exothermic reaction. When you turn the can over, heat convection will enhance the heat transfer within sake. From the information given, estimate the amount of calcium oxide and water required.

5.3   Freon Free Foam. Justify the assertion made in Example 5.3–1 that "a spoonful of NaOH" will drop the vapor pressure of CO_2 sufficiently. What do you think the rate-limiting step in this reaction is likely to be?

5.4   Reusable Cleaning Cloths. Your company wishes to make a reusable cloth to replace the use of paper towels in the kitchen. The knitted cloth features a pattern of tiny, precisely replicated peaks and troughs; the peaks collect small particles such as dust; the valleys accumulate larger particles. The fibers of the cloth are an order of magnitude thinner than a human hair to avoid smearing. The fabric is composite of polyester (hydrophilic) and nylon (hydrophobic) to allow it to absorb both oil and water. The knitting technology is important to the texture, which allows the correct performance.

(a) Which aspects of this cloth would you wish to protect by patenting?
(b) Design a series of experiments to optimize the design of the cloth, to verify its effectiveness, and to determine if the project is likely to be commercially viable.

5.5   Warm Baby Bottles. Parents of babies often wish to heat milk to around 40°C while traveling or while in locations where there is no easily available source of hot water. We propose making a device that consists of a sleeve filled with supersaturated sodium acetate trihydrate. This material crystallizes at 57°C, but spontaneous nucleation does not occur above −25°C. The product is therefore stable when molten until crystallization is initiated. This can be done by bending a cracked metal disk held in the solution. Once nucleated, heat is generated and the milk warmed.

(a) Provide final specifications for the product.

(b) Consider also the risks that might make the product unsuccessful.

5.6 **Still Better Beer.** The Guinness "widget" is a device to allow draught beer to be stored, shipped, and served in a can. It is a plastic insert filled with nitrogen, which replaces the traditional carbonated beer by nitrogen bubbles. This alters the flavor and changes the head from yellowish to a more aesthetic white.

(a) Consider what missing information the development team might have needed, once they had decided to use the plastic insert.

(b) Describe how patent protection might be gained or evaded.

5.7 **Food Wrap.** Your company plans to enter the packaging business with an oxygen impermeable food wrap. The key specification for this wrap was to keep the oxygen concentration in the food package below 0.1% for 1 year. The product chosen for development is a three layer composite consisting of 15 μm of saran, 50 μm of polypropylene, and 15 μm of low density polyethylene. The polypropylene layer contained 10 wt.% of ascorbate and 20 ppm of $CuCl_2$ catalyst, which reacts with oxygen and prevents its early permeation. Early tests with this material are very positive, so you are considering manufacturing this product.

One possible scheme for manufacture is shown in Figure P5.6. The process begins with two extruders, which make the polypropylene active layer and the polyethylene support layer. After this film is made, it is coated with saran in an emulsion coater. This saran layer is a good oxygen barrier in itself and works synergistically with the polypropylene to meet the specification.

Specify the amounts of materials and the operating parameters to produce 5000 m^2/hr of this film.

5.8 **Insulin without Needles.** One method of stabilizing proteins is to complex them with a sugar glass made of three parts sucrose and one part raffinose (*Science*, 1998,

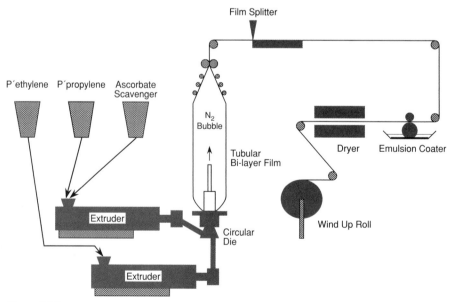

Figure P5.6.

TABLE P5.8   Markets for Isopropyl Alcohol

| | Concentration | | | |
Type	Metals	Water	Market Size($MM)	Value($/gal)
Bulk		1 wt. %		2.24
Technical grade		1000 ppm		3.75
Electronics grades	1–10 ppb	1000 ppm	6	4.90
	<1 ppb	300 ppm	12	6.80
	<100 ppb	30–50 ppm	6	11.70

281, 1793). Because such a glass inhibits protein denaturation, it can be used to keep dry proteins stable.

This glass has been suggested as the basis for an insulin aerosol which can be inhaled. Such an aerosol could allow diabetes patients to reduce the number of insulin injections that they need. This approach does not completely avoid injections: because the new form of insulin is effective only for a few hours, most patients will still need an injection to get through the night. Still, most patients will prefer inhaling insulin to a day full of needles.

Develop a process flow sheet for making such an aerosol. The process should begin with a solution, and then spray dry the solution to give particles about 3 μm in diameter. The design should include some measure of the size of the equipment, which should aim to capture 10% of the current insulin market.

5.9 **Washing Silicon Wafers.** You are interested in manufacturing ultrapure isopropyl alcohol (IPA) for cleaning microelectronic parts. These parts were formerly cleaned with freons, which were then allowed to evaporate. Because these freons damage the ozone layer in the stratosphere, they may no longer be used for this purpose. Many alternative solvents have been suggested; the one most frequently chosen for these cleaning operations is IPA.

The basic market for IPA is shown in Table P5.8. Obviously, there is a substantial premium for IPA containing low concentrations of metals and of water. You are interested in one of the four technologies shown in Figure P5.8. Choose the best, justifying your answer.

5.10 **Polypeptides to Order.** You work for a laboratory instrument company who are interested in making machines to effect chemical synthesis. As an initial target in this market, you decide to explore building a machine to make a decapeptide, a polymer of ten amino acids bound together in a specified sequence. Such a machine is sometimes called a Merrifield synthesizer.

The benchmark chosen for comparison can be idealized as a packed bed 30 cm high and 1 cm in diameter, filled with 100-μm particles. The bed uses fluoroenyl-methoxycarbonyl (Fmoc) chemistry for adding the amino acids. With this chemistry, each amino acid addition involves the following steps: start with a solid phase which already has a linker attached. Typical substitution values for PS solid phases are 1 mmol/g. Add 15 mL of reagent per gram of solid phase. Then proceed as follows:

1. Wash the column with dichloromethane (DCM) (2 × 3 min).
2. Wash with dimethylformamide (DMF) (2 × 3 min).

Technology	Schematic	Skid Cost (Capacity)	System Capabilities
Distillation	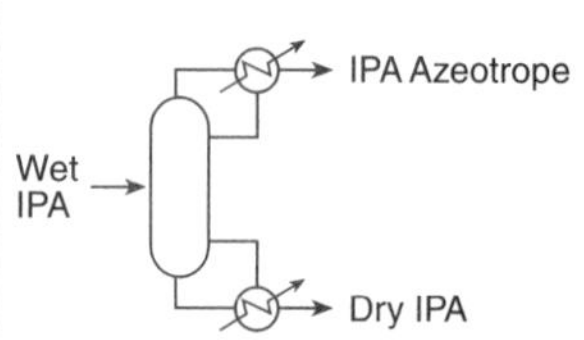	$100 K (9 gph)	Product quality — fair • 1000 ppm water • <2 ppb metals IPA recovery = 40%
Zeolite	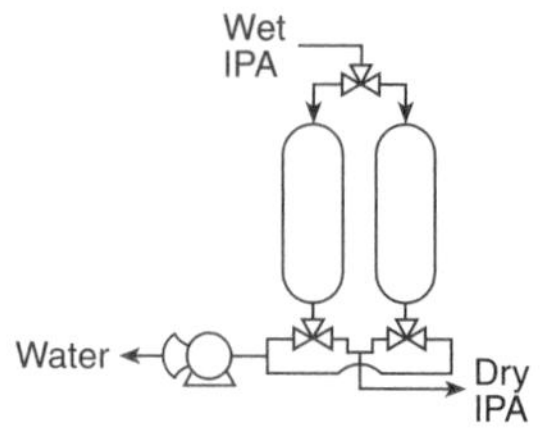	$92 K (4 gph)	Product quality — good • 100 ppm water • 3 ppb metals IPA recovery = 85%
Pervaporation	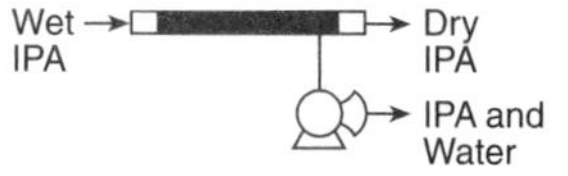	$600 K (7.5 gph)	Product quality — good • 1000 ppm water • 4 ppb metals IPA recovery = 95%
Vapor Permeation	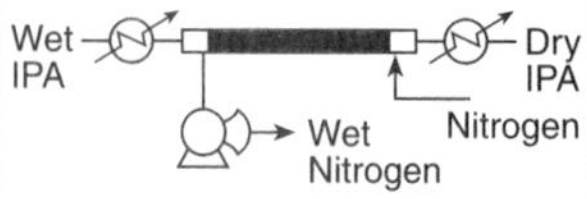	$150 K (9 gph)	Product quality — excellent • 10 ppm water • <2 ppb metals • Low particulates IPA recovery = 98%

Figure P5.8.

3. Deprotect with DMF: piperidine (1:1) (20 min).
4. Wash with DMF (2×3 min).
5. Wash with dioxane: water (2:1) (2×10 min).
6. DMF wash (3×5 min).
7. DCM wash (3×5 min).
8. Add derivitized Fmoc amino acid in DCM: DMF solution (typically 3 mol of amino acid per mole of bound linker) (30 min).
9. DCM wash (5×8 min).
10. DCF wash (5×8 min).
11. Isopropyl alcohol wash (5×8 min).
12. Repeat sequence to add the next amino acid.

The times given are for a synthesis carried out by hand, so the total time for amino acid addition is four hours. When this production is automated, as it is in commercial machines, the total time per amino acid is still one hour. Even so, the machine produces only 25 μmol of the decapeptide in a 10- to 12-hour cycle. This corresponds to perhaps 30 mg per cycle, or 60 mg per day.

Your company wants to make 100 g per day, a scale up of about 1500 times. Although you could obviously increase bed diameter $(1500)^{1/2}$ times, you would like to invent a more efficient process. Unfortunately, your efforts at generating new chemistries in new phases have not been that promising. You have been driven back to the established packed bed geometry.

Your challenge is thus to speed up the existing process. When you look at the sequence above, you are struck by the fact that three quarters of the total time is

TABLE P7.2   Financial Estimates for Zebra Mussel Control

Financial Year	Income, £	Costs, £	Profit Before Tax
2000/1	0	54,500	−54,500
2001/2	0	228,850	−228,850
2002/3	200,000	291,800	−91,800
2003/4	800,000	314,800	485,200
2004/5	2,000,000	350,800	1,649,200

washing, that is, mass transfer. You decide to accelerate this washing by using smaller particles. One quarter of the total time is chemical change, that is, reactions. You decide to accelerate these reactions with higher temperatures.

Design a new column and operating procedure based on these ideas.

CHAPTER 6. SPECIALTY CHEMICAL MANUFACTURE

6.1 Put two tablespoonfuls of salt into two similar mugs. Pour over just sufficient boiling water to dissolve the salt. Place one mug in the freezer until cool; leave the other to cool more slowly in the room. Decant off the water and observe the salt crystals that have formed. Explain the disparity.

6.2 How would you separate a mixture of salt and pepper in the kitchen? Try it.

6.3 Estimate the time saved by cooking rice in a pressure cooker which uses 40 kPa overpressure. Is this a heat transfer or reaction rate-limited problem? Do some experiments to support your answer. Why are pressure cookers used at high altitude?

6.4 Carry out the plot suggested in Section 6.2–2. Suggest possible problems with this analysis. Are they likely to be serious?

6.5 Discuss why stringent rules exist in pharmaceuticals manufacture, giving examples.

6.6 Suggest a protocol for separating and purifying insulin from a broth containing genetically modified bacteria, which produce the drug.

CHAPTER 7. ECONOMIC CONCERNS

7.1 Your company proposes to develop a new range of vacuum cleaners by using proprietary technology. If the development costs are expected to be $1,000,000 over two years, how many vacuum cleaners will need to be sold at $250 for the project to be financially attractive over a five year period?

(a) Ignore the time value of money.
(b) Include the cost of money at 16% annually.

7.2 One measure often used in evaluating a product design project is the internal rate of return (IRR). The IRR is the annual interest rate (cost of money) at which the net present value of the project would become zero. A proposed solution to the zebra mussel problem outlined in Chapter 5 (Example 5.2–3) has the projected profit and loss in Table P7.2. Calculate the IRR over this five year period. Then discuss.

What type of assumptions are likely to have been made in producing these estimates of income and costs? List the key risks you would like to evaluate further before putting your money into this project.

Index

Products Index